Oxford Revise

Revision & Practice

AQA GCSE 9–1 BIOLOGY HIGHER

 **Knowledge**  **Retrieval**  **Practice**

Series Editor: **Primrose Kitten**

Jo Locke
Jessica Walmsley

OXFORD
UNIVERSITY PRESS

 Retrieval  Practice

confident and ready for your exam.

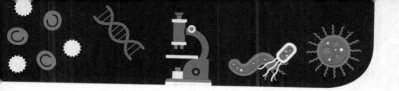

Paper 2 starts here

use this book

This book uses a three-step approach to revision: **Knowledge**, **Retrieval**, and **Practice**.
It is important that you do all three; they work together to make your revision effective.

1 Knowledge

Knowledge comes first. Each chapter starts with a **Knowledge Organiser**. These are clear, easy-to-understand, concise summaries of the content that you need to know for your exam. The information is organised to show how one idea flows into the next so you can learn how all the science is tied together, rather than lots of disconnected facts.

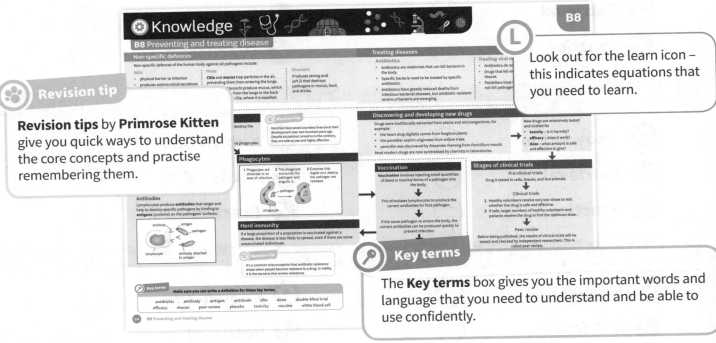

Revision tip

Revision tips by **Primrose Kitten** give you quick ways to understand the core concepts and practise remembering them.

Look out for the learn icon – this indicates equations that you need to learn.

Key terms

The **Key terms** box gives you the important words and language that you need to understand and be able to use confidently.

2 Retrieval

The **Retrieval questions** help you learn and quickly recall the information you've acquired. These are short questions and answers about the content in the Knowledge Organiser. Cover up the answers with some paper; write down as many answers as you can from memory. Check back to the Knowledge Organiser for any you got wrong, then cover the answers and attempt *all* the questions again until you can answer all the questions correctly.

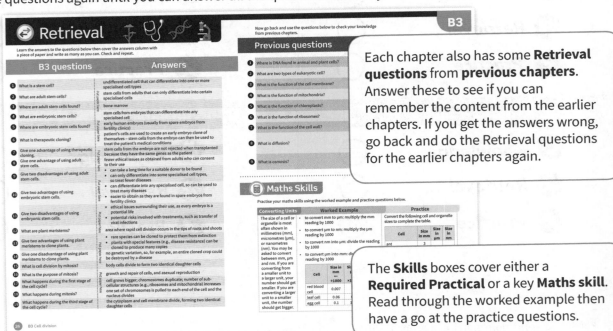

Each chapter also has some **Retrieval questions** from **previous chapters**. Answer these to see if you can remember the content from the earlier chapters. If you get the answers wrong, go back and do the Retrieval questions for the earlier chapters again.

The **Skills** boxes cover either a **Required Practical** or a key **Maths skill**. Read through the worked example then have a go at the practice questions.

Make sure you revisit the retrieval questions on different days to help them stick in your memory. You need to write down the answers each time, or say them out loud, otherwise it won't work.

3 Practice

Once you think you know the Knowledge Organiser and Retrieval answers really well you can move on to the final stage: **Practice**.

Each chapter has lots of **exam-style questions**, including some questions from previous chapters, to help you apply all the knowledge you have learnt and can retrieve.

Each question has a difficulty icon that shows the level of challenge.

 These questions build your confidence.

 These questions consolidate your knowledge.

 These questions stretch your understanding.

Make sure you attempt all of the questions no matter what grade you are aiming for.

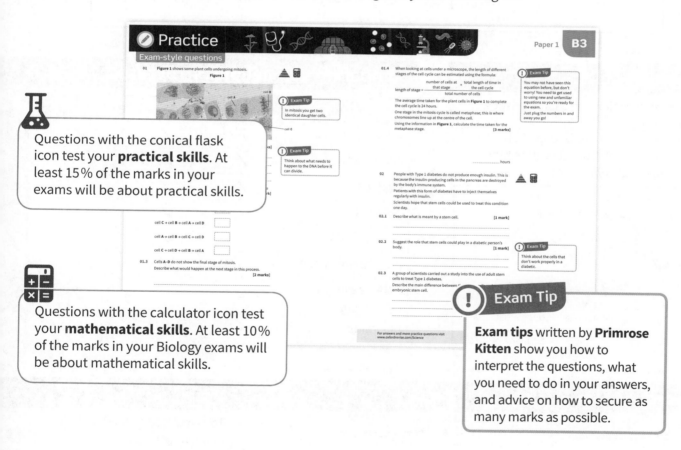

Questions with the conical flask icon test your **practical skills**. At least 15% of the marks in your exams will be about practical skills.

Questions with the calculator icon test your **mathematical skills**. At least 10% of the marks in your Biology exams will be about mathematical skills.

Exam Tip

Exam tips written by **Primrose Kitten** show you how to interpret the questions, what you need to do in your answers, and advice on how to secure as many marks as possible.

kerboodle

All the **answers** are on Kerboodle and the website, along with even more exam-style questions. www.oxfordrevise.com/scienceanswers

B1 Cell biology

Eukaryotic cells

Animal and plant cells are eukaryotic cells. They have genetic material (DNA) that forms **chromosomes** and is contained in a **nucleus**.

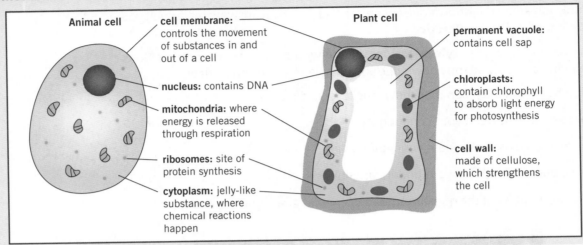

Animal cell

cell membrane: controls the movement of substances in and out of a cell

nucleus: contains DNA

mitochondria: where energy is released through respiration

ribosomes: site of protein synthesis

cytoplasm: jelly-like substance, where chemical reactions happen

Plant cell

permanent vacuole: contains cell sap

chloroplasts: contain chlorophyll to absorb light energy for photosynthesis

cell wall: made of cellulose, which strengthens the cell

Prokaryotic cells

Bacteria have the following characteristics:
- single-celled
- no nucleus – have a single loop of DNA
- have small rings of DNA called **plasmids**
- smaller than eukaryotic cells.

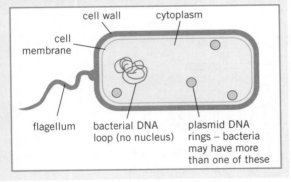

cell wall   cytoplasm

cell membrane

flagellum   bacterial DNA loop (no nucleus)   plasmid DNA rings – bacteria may have more than one of these

Comparing sub-cellular structures

Structure	Animal cell	Plant cell	Prokaryotic cell
cell membrane	✓	✓	✓
cytoplasm	✓	✓	✓
nucleus	✓	✓	—
cell wall	—	✓	✓
chloroplasts	—	✓	—
permanent vacuole	—	✓	—
DNA free in cytoplasm	—	—	✓
plasmids	—	—	✓

Microscopes

Light microscope	Electron microscope
uses light to form images	uses a beam of electrons to form images
living samples can be viewed	samples cannot be living
relatively cheap	expensive
low magnification	high magnification
low resolution	high resolution

Electron microscopes allow you to see sub-cellular structures, such as ribosomes, that are too small to be seen with a light microscope.

To calculate the magnification of an image:

Ⓛ $\text{magnification} = \dfrac{\text{image size}}{\text{actual size}}$

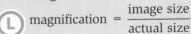

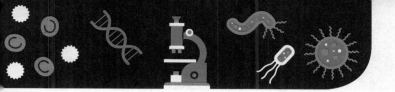

Specialised cells

Cells in animals and plants differentiate to form different types of cells. Most animal cells differentiate at an early stage of development, whereas a plant's cells differentiate throughout its lifetime.

Specialised cell	Function	Adaptations
sperm cell	fertilise an ovum (egg)	• tail to swim to the ovum and fertilise it • lots of mitochondria to release energy from respiration, enabling the sperm to swim to the ovum
red blood cell	transport oxygen around the body	• no nucleus so more room to carry oxygen • contains a red pigment called haemoglobin that binds to oxygen molecules • flat bi-concave disc shape to increase surface area-to-volume ratio
muscle cell	contract and relax to allow movement	• contains protein fibres, which can contract to make the cells shorter • contains lots of mitochondria to release energy from respiration, allowing the muscles to contract
nerve cell	carry electrical impulses around the body	• branched endings, called dendrites, to make connections with other neurones or effectors • myelin sheath insulates the axon to increase the transmission speed of the electrical impulses
root hair cell	absorb mineral ions and water from the soil	• long projection speeds up the absorption of water and mineral ions by increasing the surface area of the cell • lots of mitochondria to release energy for the active transport of mineral ions from the soil
palisade cell	enable photosynthesis in the leaf	• lots of chloroplasts containing chlorophyll to absorb light energy • located at the top surface of the leaf where it can absorb the most light energy

🔑 **Key terms**

Make sure you can write a definition for these key terms.

cell membrane	cell wall	chloroplast	chromosome	cytoplasm	DNA
eukaryotic	magnification	mitochondria	nucleus	plasmid	
prokaryotic	resolution	ribosome	permanent vacuole		

Learn the answers to the questions below, then cover the answers column with
a piece of paper and write as many as you can. Check and repeat.

B1 questions | Answers

#	Question	Answer
1	What are two types of eukaryotic cell?	animal and plant
2	What type of cell are bacteria?	prokaryotic
3	Where is DNA found in animal and plant cells?	in the nucleus
4	What is the function of the cell membrane?	controls movement of substances in and out of the cell
5	What is the function of mitochondria?	site of respiration to transfer energy for the cell
6	What is the function of chloroplasts?	contain chlorophyll to absorb light energy for photosynthesis
7	What is the function of ribosomes?	enable production of proteins (protein synthesis)
8	What is the function of the cell wall?	strengthens and supports the cell
9	What is the structure of the main genetic material in a prokaryotic cell?	single loop of DNA
10	How are electron microscopes different to light microscopes?	electron microscopes use beams of electrons instead of light, cannot be used to view living samples, are much more expensive, and have a much higher magnification and resolution
11	What is the function of a red blood cell?	carries oxygen around the body
12	Give three adaptations of a red blood cell.	no nucleus, contains a red pigment called haemoglobin, and has a bi-concave disc shape
13	What is the function of a nerve cell?	carries electrical impulses around the body
14	Give two adaptations of a nerve cell.	branched endings, myelin sheath insulates the axon
15	What is the function of a sperm cell?	fertilises an ovum (egg)
16	Give two adaptations of a sperm cell.	tail, contains lots of mitochondria
17	What is the function of a palisade cell?	carries out photosynthesis in a leaf
18	Give two adaptations of a palisade cell.	lots of chloroplasts, located at the top surface of the leaf
19	What is the function of a root hair cell?	absorbs minerals and water from the soil
20	Give two adaptations of a root hair cell.	long projection, lots of mitochondria

Put paper here

Maths Skills

Practise your maths skills using the worked example and practice questions below.

Resolution	Worked Example	Practice
The resolution of a device is the smallest change that the device can measure. Selecting equipment with the appropriate resolution is important in scientific investigations. If the resolution of a digital watch is one second, one second is the smallest amount of time it can measure. Some stop clocks have smaller resolutions, for example a resolution of 0.01 seconds. This means that they can measure times of 0.01, 1.29 or 9.62 seconds, whereas a digital watch could not.	What is the resolution of the following equipment?  The resolution of this thermometer is 1 °C, as this is the smallest change that it can detect.  The resolution of this digital thermometer is 0.1 °C, as it can measure readings such as 1.1 °C, 8.9 °C and 36.7 °C.	What are the resolutions of the following pieces of equipment? 1  2  3 

Required Practical Skills

Practise answering questions on the required practicals using the example below. You need to be able to apply your skills and knowledge to other practicals too.

Looking at cells	Worked example	Practice
In this practical you need to be able to use a light microscope to view plant and animal cells. You should be able to: • describe how to set up a microscope • label parts of a microscope • describe how to focus on a slide containing a specimen • make a labelled scientific drawing of what you observe. You also need to be able to determine the magnification of an object under a microscope, and use this to calculate the real size of the object.	A student wanted to determine the actual size of the cell they observed under a microscope. They measured the size of the cell as 15 mm, the objective lens magnification was 40×, and the eyepiece magnification was 10×. Determine the actual size of the cell. Give your answer in standard form. **Step 1:** determine the magnification $$\text{total magnification} = \text{objective lens magnification} \times \text{eyepiece lens magnification}$$ total magnification = 40 × 10 = 400 **Step 2:** put the numbers in the equation $$\text{magnification} = \frac{\text{size of image}}{\text{actual size of object}}$$ $$400 = \frac{15}{\text{actual size of object}}$$ **Step 3:** rearrange the equation and find the answer $$\text{actual size of object} = \frac{15}{400} = 0.0375 \text{ mm}$$ **Step 4:** convert your answer to standard form $0.0375 = 3.75 \times 10^{-2}$ mm	1 Describe how you could identify a cell as animal or plant by looking at it using a light microscope. Include the names and visual descriptions of any important organelles. 2 A student wrote a method for focusing a microscope image of a cell. Suggest improvements to the student's method. "Set the microscope at the highest objective lens, then look down the eyepiece while you use the fine focus to find the cell." 3 Draw a labelled image of an animal cell.

 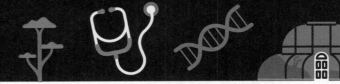

Practice

Exam-style questions

01 **Figure 1** shows a plant cell.

Figure 1

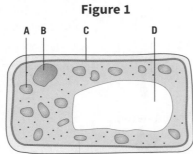

01.1 Identify parts **A–C**. **[3 marks]**

A _____

B _____

C _____

> **! Exam Tip**
>
> Labelling cells is a really common exam question. Take the time to become familiar with plant cells, animal cells, and prokaryotic cells.

01.2 Explain the function of part **D**. **[3 marks]**

_____

_____

_____

_____

01.3 Which feature in **Figure 1** allows you to conclude that this is a eukaryotic cell? **[1 mark]**

Tick **one** box.

cell wall ☐

nucleus ☐

cytoplasm ☐

cell membrane ☐

01.4 The cell shown in **Figure 1** was taken from a plant.

Suggest and explain where in the plant this cell would be found. **[2 marks]**

_____

_____

_____

> **! Exam Tip**
>
> Plants have a range of different specialised cells, but they will all share some common features.

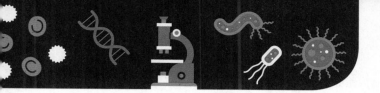 

02 **Figure 2** shows a cheek cell viewed under a light microscope magnified at ×1350.

Figure 2

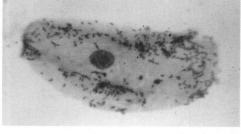

02.1 A student collected a sample of cells by taking a saliva swab on the inside of their cheek.

Explain **one** safety measure that the student should take during this procedure. **[2 marks]**

_____

_____

_____

> **! Exam Tip**
>
> For this question 'explain' is the command word so you need to say *why* you have done something, not just what you have done.

02.2 Explain why the dye methylene blue was added to the cell sample on the slide. **[1 mark]**

_____

_____

02.3 Suggest **one** way in which the student could observe structures within the cell in greater detail. **[1 mark]**

_____

_____

02.4 Using **Figure 2**, calculate the actual length of the cheek cell.

Give your answer in micrometres. **[3 marks]**

> **! Exam Tip**
>
> Converting between units can be tricky. If you're not sure whether you need to divide or multiply, try to think logically and ask yourself:
> - if 10 mm = 1 cm, does the answer look correct if I multiply or divide?

_____ μm

03 Plant cells have two main types of transport tissue.

03.1 Describe what is meant by a tissue. **[1 mark]**

03.2 Xylem tissue transports water and mineral ions around the plant. Explain how the xylem tubes form. **[3 marks]**

> **! Exam Tip**
>
> A quick way to remember what xylem carries is that x and w (for water) are close to each other in the alphabet.

03.3 Explain **two** ways xylem tissue is adapted to its function. **[4 marks]**

03.4 Name the other type of plant transport tissue. **[1 mark]**

04   **Figure 3** shows a single-celled organism called *Euglena*.

Euglena are found in ponds and lakes.

They survive by making their own food through photosynthesis.

In low light conditions they can engulf other microorganisms, such as bacteria and algae.

Figure 3

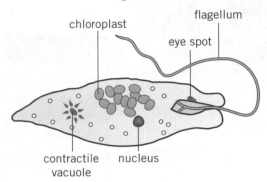

> **! Exam Tip**
>
> You may not have come across *Euglena* before. Don't let this worry you! The exam has to include a lot of content you haven't seen before, so questions like this are designed to prepare you for that.
>
> Use the knowledge you have gained in class and practise applying it to this new situation. The more you try doing this, the better prepared you'll be for the unfamiliar contexts you will come across in the exams!

04.1   Identify which part of the *Euglena* traps light for photosynthesis.   **[1 mark]**

04.2   The eye spot on the *Euglena* detects light.

Suggest how the flagellum and eye spot work together to maximise photosynthesis.   **[3 marks]**

05   Amoebas are single-celled organisms.

Naegleria fowleri, often referred to as the 'brain-eating amoeba', is a species of amoeba that can cause sudden and severe brain infection, which normally results in fatality.

05.1   Evaluate the advantages and disadvantages of using light microscopes and electron microscopes to study amoebas. **[6 marks]**

05.2   *Naegleria* measure approximately 10 µm in diameter.

The diameter of a human egg cell is approximately 0.1 mm.

Calculate the difference in orders of magnitude between *Naegleria* and a human egg cell.   **[3 marks]**

> **! Exam Tip**
>
> This is an 'evaulate' question. To get full marks you need to have a balanced argument and give a reasoned opinion. Try this plan:
> 1 advantages of light microscopes
> 2 disadvantages of light microscopes
> 3 advantages of electron microscopes
> 4 disadvantages of electron microscopes
> 5 your opinion on which one is best for the task
> 6 give a reason why you think this.

06   **Figure 4** shows some plant cells as viewed under a light microscope.

06.1   Identify the cell membrane in **Figure 4**.   **[1 mark]**

06.2   Ribosomes are present in plant cells but cannot be seen using a light microscope. Describe the function of ribosomes.

[1 mark]

Figure 4

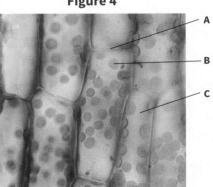

06.3 Name **one** other subcellular structure that is present in plant cells but cannot be seen in **Figure 4**. **[1 mark]**

06.4 Describe how to prepare and view a sample of plant cells using a light microscope.

[6 marks]

07 Bacteria are an example of prokaryotes.

07.1 Which of the following is the most approximate size for a prokaryote?

Choose **one** answer. **[1 mark]**

100 nm   1 μm   10 μm   0.1 mm

07.2 Both plant cells and prokaryotic cells have cell walls. Describe **one** difference between the cell wall of a plant cell, and the cell wall of a bacterial cell. **[1 mark]**

07.3 Describe the differences between the way genetic material is stored in a prokaryotic cell and in a eukaryotic cell. **[4 marks]**

07.4 Suggest which feature needs to be present on a bacterial cell if it needs to move in water. **[1 mark]**

> **! Exam Tip**
>
> Very large and very small units can seem more complicated than they really are. Make sure you know the order from largest to smallest:
>
> km > m > cm > mm > μm > nm > pm

08 Cone cells are a type of cell found at the back of the human eye.

They detect light and send information to the brain, which it then decodes, allowing us to perceive colour.

Figure 5 shows the main adaptations of a cone cell.

Figure 5

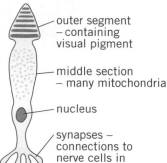

outer segment – containing visual pigment

middle section – many mitochondria

nucleus

synapses – connections to nerve cells in optic nerve

08.1 Use the information in **Figure 5** and your own knowledge to suggest and explain how the cone cell is adapted to its function.

[4 marks]

> **! Exam Tip**
>
> All the information you need is given to you in the question. Make sure that in your answer you go through each of the adaptations one by one.

08.2 There are approximately 6 million cone cells in the human retina.

Three different types of cone cell exist:

- one to detect red light
- one to detect green light
- one to detect blue light.

Assuming there is a roughly equal number of each cone cell type in a human retina, calculate the number of 'red' cone cells present in the retina.

Give your answer in standard form. **[3 marks]**

> **! Exam Tip**
>
> Read the question carefully; if you don't give your answer in standard form you are not going to get full marks.

09   Muscle cells are an example of a specialised cell.

09.1   Define the term specialised cell.   **[1 mark]**

09.2   The biceps contain muscle cells. Describe the function of a muscle cell.   **[1 mark]**

09.3   In addition to providing movement to the skeleton, muscle tissue has other functions in the body. Describe **one** other example of where muscles are found in the body.   **[2 marks]**

09.4   Explain why muscle cells have lots of mitochondria.   **[2 marks]**

09.5   Explain **one** other feature of a muscle cell.   **[2 marks]**

10   A student observed some onion cells under a microscope.

10.1   Give **two** features that help the student to know that they are looking at a sample of plant cells in **Figure 6**.   **[2 marks]**

Figure 6

10.2   The scale ruler in the diagram represents 1000 µm. Calculate the average width of an onion cell.   **[2 marks]**

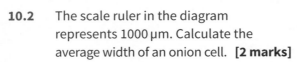

10.3   These cells were observed under 100× magnification. Which of the following structures may have been visible under a higher magnification using a light microscope? Choose **one** answer.   **[1 mark]**

> **! Exam Tip**
>
> When measuring very small things, it is better to measure a group of them and then divide the measurement, for example, measure the width of ten onion cells then divide that measurement by ten.

vacuole   ribosomes   chloroplasts   plasmids

11   **Figure 7** shows an animal cell.

Figure 7

11.1   Which letter represents the cell membrane?   **[1 mark]**

11.2   Describe the function of the cell membrane.   **[1 mark]**

11.3   **Figure 7** shows a human skin cell. Explain how the cell would differ if it was a human nerve cell.   **[4 marks]**

> **! Exam Tip**
>
> Think about the function of nerve cells. Give the difference and then explain *why* this difference is important.

11.4   An animal cell is measured to have a mean diameter of 20 µm. Estimate the length of the cell membrane material in this cell.   **[2 marks]**

11.5   The mean length of the molecules within the cell membrane has been estimated to be 4 nm. If we assume that the cell membrane is one molecule thick, calculate the total number of molecules contained in the cell's membrane.   **[3 marks]**

12 In general eukaryotic cells are one order of magnitude larger than prokaryotic cells.

Figure 8

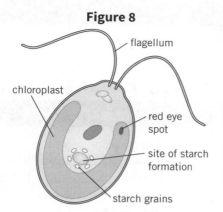

flagellum

chloroplast

red eye spot

site of starch formation

starch grains

12.1 One type of virus is 100 nm in height. One type of algal cell is two orders of magnitude larger than the virus. Estimate the length of the eukaryotic cell. Give your answer in micrometres. **[1 mark]**

12.2 Compare the features found in eukaryotic and prokaryotic cells. **[6 marks]**

12.3 Cyanobacteria are a phylum of bacteria that are able to photosynthesise. Estimate the size of a cyanobacterial cell. Justify your answer. **[2 marks]**

12.4 Algae are aquatic organisms that belong to the kingdom Protista. **Figure 8** shows an algal cell.

Use **Figure 8** to suggest why some algae have been classified as plants. **[3 marks]**

13 Sperm cells are one type of sex cell.

13.1 Explain how a sperm cell is adapted to perform its function. **[4 marks]**

13.2 Low sperm mobility is one cause of fertility issues in males. If a male is having fertility issues, a sample of sperm can be observed through a microscope to check the mobility of the sperm. Suggest **two** reasons why the sample is normally observed using a light microscope, rather than using an electron microscope. **[2 marks]**

13.3 A scientist views an image of a sperm cell. The image measures 7.5 cm, and has been magnified 1500×. Calculate the real size of the sperm cell. Give your answer in micrometres. **[3 marks]**

14 A scientist produced an image of human cheek cells. It was prepared using a light microscope.

14.1 Describe how to prepare a slide of cheek cells to view under a light microscope. **[6 marks]**

14.2 Sketch a diagram of the cell you would expect to see. Label the organelles which are visible in the cell. **[3 marks]**

14.3 Write down which additional piece of information you should include with a microscope drawing. **[1 mark]**

14.4 Some organelles will not be visible because a light microscope does not have a high enough resolution. Define the term resolution. **[1 mark]**

14.5 Suggest how the scientist could produce an image showing the missing organelles. **[1 mark]**

B2 Cell transport

Comparing diffusion, osmosis, and active transport

	Diffusion	Osmosis	Active transport
Definition	The spreading out of particles, resulting in a net movement from an area of higher **concentration** to an area of lower concentration.	The diffusion of water from a **dilute** solution to a concentrated solution through a **partially permeable membrane**.	The movement of particles from a more dilute solution to a more concentrated solution using energy from respiration.
Movement of particles	Particles move down the concentration **gradient** – from an area of *high* concentration to an area of *low* concentration.	Water moves from an area of *lower* solute concentration to an area of *higher* solute concentration.	Particles move against the concentration gradient – from an area of *low* concentration to an area of *high* concentration.
Energy required?	no – **passive process**	no – passive process	yes – energy released by respiration
Examples	**Humans** • Nutrients in the small intestine diffuse into the **capillaries** through the **villi**. • Oxygen diffuses from the air in the **alveoli** into the blood in the capillaries. Carbon dioxide diffuses from the blood in the capillaries into the air in the alveoli. • **Urea** diffuses from cells into the blood for excretion in the kidney. **Fish** • Oxygen from water passing over the gills diffuses into the blood in the **gill filaments**. • Carbon dioxide diffuses from the blood in the gill filaments into the water. **Plants** • Carbon dioxide used for photosynthesis diffuses into leaves through the **stomata**. • Oxygen produced during photosynthesis diffuses out of the leaves through the stomata.	**Plants** Water moves by osmosis from a dilute solution in the soil to a concentrated solution in the **root hair cell**.	**Humans** Active transport allows sugar molecules to be absorbed from the small intestine when the sugar concentration is higher in the blood than in the small intestine. **Plants** Active transport is used to absorb mineral ions into the root hair cells from more dilute solutions in the soil.

diffusion — particles move **down the concentration gradient**

high concentration → low concentration

osmosis — water molecules move **from a dilute to a concentrated solution**

dilute solution — water molecule — partially permeable membrane — solute particle — concentrated solution

active transport — particles move **against the concentration gradient** + energy

low concentration → high concentration

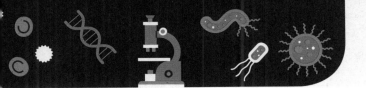

Factors that affect the rate of diffusion

① **Difference in concentration**

The steeper the concentration gradient, the faster the rate of diffusion.

② **Temperature**

The higher the temperature, the faster the rate of diffusion.

③ **Surface area of the membrane**

The larger the membrane surface area, the faster the rate of diffusion.

Adaptations for exchanging substances

Single-celled organisms have a large surface area-to-volume ratio. This means enough molecules can be transported across their cell membranes to meet their needs.

Multicellular organisms have a small surface area-to-volume ratio. This means they need specialised organ systems and cells to allow enough molecules to be transported into and out of their cells.

Exchange surfaces work most efficiently when they have a large surface area, a thin membrane, and a good blood supply.

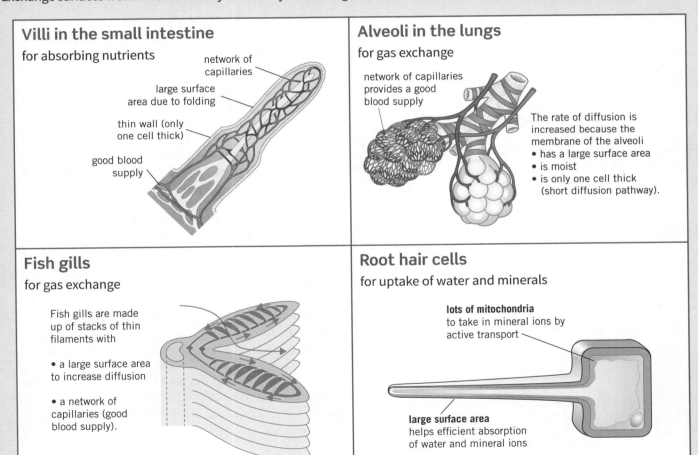

Villi in the small intestine
for absorbing nutrients

network of capillaries

large surface area due to folding

thin wall (only one cell thick)

good blood supply

Alveoli in the lungs
for gas exchange

network of capillaries provides a good blood supply

The rate of diffusion is increased because the membrane of the alveoli
- has a large surface area
- is moist
- is only one cell thick (short diffusion pathway).

Fish gills
for gas exchange

Fish gills are made up of stacks of thin filaments with

- a large surface area to increase diffusion

- a network of capillaries (good blood supply).

Root hair cells
for uptake of water and minerals

lots of mitochondria to take in mineral ions by active transport

large surface area helps efficient absorption of water and mineral ions

🔑 **Key terms**

Make sure you can write a definition for these key terms.

active transport   alveoli   capillaries   concentration   diffusion   dilute
gill filament   gradient   osmosis   partially permeable membrane
passive process   root hair cell   stomata   urea   villi

Retrieval

Learn the answers to the questions below then cover the answers column with
a piece of paper and write as many as you can. Check and repeat.

B2 questions	Answers
1 What is diffusion?	net movement of particles from an area of high concentration to an area of low concentration along a concentration gradient – this is a passive process (does not require energy from respiration)
2 Name three factors that affect the rate of diffusion.	concentration gradient, temperature, membrane surface area
3 How are villi adapted for exchanging substances?	• long and thin – increases surface area • one-cell-thick membrane – short diffusion pathway • good blood supply – maintains a steep concentration gradient
4 How are the lungs adapted for efficient gas exchange?	• alveoli – large surface area • moist membranes – increases rate of diffusion • one-cell-thick membranes – short diffusion pathway • good blood supply – maintains a steep concentration gradient
5 How are fish gills adapted for efficient gas exchange?	• large surface area for gases to diffuse across • thin layer of cells – short diffusion pathway • good blood supply – maintains a steep concentration gradient
6 What is osmosis?	diffusion of water from a dilute solution to a concentrated solution through a partially permeable membrane
7 Give one example of osmosis in a plant.	water moves from the soil into the root hair cell
8 What is active transport?	movement of particles against a concentration gradient – from a dilute solution to a more concentrated solution – using energy from respiration
9 Why is active transport needed in plant roots?	concentration of mineral ions in the soil is lower than inside the root hair cells – the mineral ions must move against the concentration gradient to enter the root hair cells
10 What is the purpose of active transport in the small intestine?	sugars can be absorbed when the concentration of sugar in the small intestine is lower than the concentration of sugar in the blood

Put paper here

Now go back and use the questions below to check your knowledge from previous chapters.

Previous questions

Answers

Previous questions		Answers
Give two adaptations of a root hair cell.	*Put paper here*	long projection, lots of mitochondria
What is the function of a red blood cell?		carries oxygen around the body
What type of cell are bacteria?		prokaryotic
What is the function of ribosomes?	*Put paper here*	enable production of proteins (protein synthesis)
Give two adaptations of a nerve cell.		branched endings, myelin sheath insulates the axon
What is the function of a sperm cell?	*Put paper here*	fertilises an ovum (egg)
Give two adaptations of a sperm cell.		tail, contains lots of mitochondria
How are electron microscopes different to light microscopes?		electron microscopes use beams of electrons instead of light, cannot be used to view living samples, are much more expensive, and have a much higher magnification and resolution

Required Practical Skills

Practise answering questions on the required practicals using the example below. You need to be able to apply your skills and knowledge to other practicals too.

Osmosis in cells	Worked example	Practice
Different concentrations of sugar and salt solutions both affect the movement of water by osmosis, causing cells to lose or gain water and changing the mass of a tissue sample. For this practical you need to be able to accurately measure length, mass, and volume to measure osmosis in cells. You will need to be comfortable applying this knowledge to a range of samples, not just to the typical example of potato tissue, as osmosis happens in all cells.	A sample of carrot was placed into a 0.75 mol/dm³ sugar solution for 30 minutes. The mass of the carrot was recorded before and after this. Initial mass = 6.02 g Final mass = 3.91 g 1 Determine the percentage change in mass of the sample. $$3.91 - 6.02 = -2.11\,g$$ $$\left(\frac{-2.11}{6.02}\right) \times 100 = -35\,\%$$ (a minus sign is used because the sample has lost mass) 2 Explain why this experiment should be repeated, and give one other variable that should be controlled. The experiment should be repeated to give a more reliable result, and to allow calculation of a mean loss in mass for the sample. The dimensions of the carrot sample needs to be controlled between repeats.	1 Give one reason why it is important to dry the samples of carrot cores before they are weighed. 2 When repeating this experiment using different concentrations of sugar solution, a student found that one sample did not change mass. Suggest what this tells you about the concentration of the solution. Assume no error in the experiment. 3 Two students set up this experiment. Student A said that each sample of carrot must have the same starting mass. Student B argued that each sample must have the same length and width. Explain which student is correct.

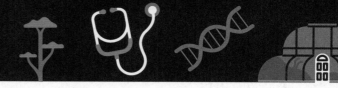
01 A group of students investigated how the mass of a potato sample changed over time, when placed into sugar solutions of varying concentrations.

They set up their equipment as shown in **Figure 1**.

Figure 1

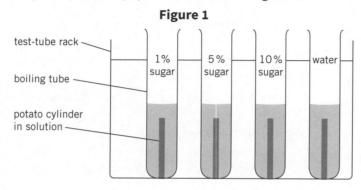

test-tube rack

boiling tube

potato cylinder in solution

1% sugar   5% sugar   10% sugar   water

01.1 Name the independent variable in their investigation. **[1 mark]**

_____

01.2 Identify **two** variables that the students controlled. **[2 marks]**

1 _____

2 _____

> ⚠ **Exam Tip**
>
> Control variable are the ones we keep the same.

01.3 The students' results are shown in **Table 1**.

Table 1

Percentage sugar solution	0%	1%	5%	10%
Starting mass in grams	3.2	3.3	3.1	3.4
Final mass in grams	3.7	3.5	2.9	2.6
Change in mass in grams	+0.5	_____	−0.2	−0.8
Percentage change in mass	+15.6	_____	−6.5	−23.5

> ⚠ **Exam Tip**
>
> If you're not sure what to do try using the values given for the 5% and 10% solutions as trials, and see if you can get the answer.

Complete the results table by calculating the change in mass and the percentage change in mass for the 1% sugar solution. **[2 marks]**

_____

_____

_____

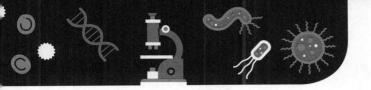
01.4 Plot the students' results of sugar concentration against percentage change in mass on the axes in **Figure 2**.

Draw a suitable line of best fit. **[3 marks]**

Figure 2

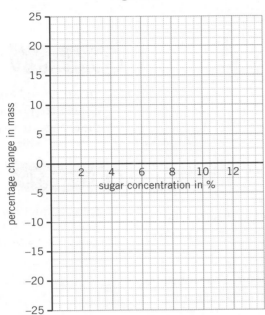

01.5 Determine the concentration of sugar present in the potato. **[1 mark]**

_____

01.6 Describe what the students should do to check that their results are reproducible. **[2 marks]**

_____

_____

_____

02 Multicellular organisms often have complex structures, such as lungs, for exchanging gases.

Explain why single-celled organisms, like *Euglena*, do not need lungs. **[3 marks]**

_____

_____

_____

_____

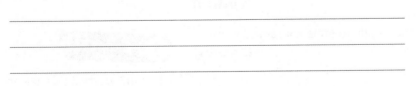

03   **Figure 3** shows a plant cell before (**A**) and after (**B**) it was placed in a solution of salt and water.

Figure 3

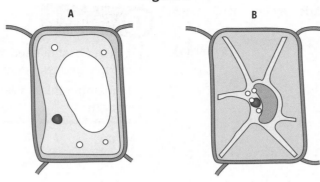

03.1   Name the type of solution caused the differences observed in diagram **B**.   **[1 mark]**

03.2   Explain the differences in the appearance of the plant cell after it was placed in the saltwater solution.   **[6 marks]**

03.3   Explain why the cells in **Figure 3** can only be seen in this state in a laboratory.   **[1 mark]**

04   **Figure 4** shows a number of different ways substances can move into and out of a cell. The dots represent the molecules of each substance.

Figure 4

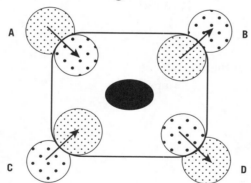

04.1   Name the cell structure that controls the movement of materials into and out of the cell.   **[1 mark]**

04.2   Identify which arrow represents the active uptake of glucose by epithelial cells in the small intestine. Give a reason for your answer.   **[2 marks]**

04.3   Explain why epithelial cells in the small intestine contain so many mitochondria.   **[3 marks]**

> **! Exam Tip**
>
> The command word here is *explain*, so you need to saw WHY the changes have happened.

> **! Exam Tip**
>
> Think about how we visualise cells.

> **! Exam Tip**
>
> Start this question by labelling the high and low concentrations.

> **! Exam Tip**
>
> Don't worry if this is a new context for you – in the exam you'll have to apply what you have learnt in class to new and unfamiliar things.

05 Gas exchange in fish takes place in the gills (**Figure 5**). Fish breathe by taking in oxygen from their environment through opening their mouths underwater. This allows water, containing oxygen, to pass over their gills, causing the oxygen to pass from the water into the fish's bloodstream.

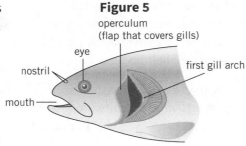

Figure 5

05.1 Describe what is meant by gas exchange. **[1 mark]**

05.2 Use **Figure 5** to suggest **two** ways in which fish gills are adapted for efficient gas exchange. **[2 marks]**

05.3 Students investigated the effect of temperature on the breathing rate of fish. They put same-sized fish in tanks of water at five different temperatures. They then measured the breathing rate of the fish by counting the number of times the fish opened their mouths in one minute.

The fish were placed in the water tanks for five minutes before the investigation began.

Suggest why the students included this step in their investigation. **[1 mark]**

> ! **Exam Tip**
>
> Think about jumping into the swimming pool for the first time – does the temperature feel the same then as after 5 minutes?

05.4 The students repeated the test five times at each temperature then calculated the mean. Their results are shown in **Table 2**.

Table 2

Temperature of water in °C	Mean number of breaths taken per minute
5	20
10	38
15	45
20	57
25	70

Describe the trend shown in the results. **[1 mark]**

> ! **Exam Tip**
>
> This is a 1 mark *describe* question, so it's looking for a short answer stating what the pattern is.

05.5 Identify which conclusion the students could draw from their results. Choose **one**. **[1 mark]**

The oxygen content of water remains the same regardless of temperature.

The oxygen content of water increases as temperature increases.

The oxygen content of water decreases as temperature increases.

06 *Crocodylus porosus* is a species of freshwater crocodile which normally lives in lakes and rivers. This species of crocodile is also able to survive in salt water because they have special salt glands in their tongues that remove excess salt from their bodies.

06.1 Explain why *Crocodylus porosus* have to use active transport to remove excess salt from their body when living in a saltwater environment. **[2 marks]**

06.2 Explain why the cells in the salt glands have large numbers of mitochondria. **[3 marks]**

06.3 Suggest **one** advantage to *Crocodylus porosus* of being able to inhabit both saltwater and freshwater habitats. **[1 mark]**

 Exam Tip

Long Latin names can be confusing. I used to get my students to cross out the long confusing name and replace it with something friendly – just don't use the friendly name in your exam answer!

07 A scientist planned to carry out an investigation to determine which variety of apple was the sweetest. The scientist had access to the following equipment:

- range of different apples
- potato borer
- scalpel
- balance
- distilled water
- test tubes
- test-tube rack
- measuring cylinder
- sucrose solutions at six different concentrations.

Plan an investigation the scientist could follow to determine the sugar concentration of each variety of apple. **[6 marks]**

 Exam Tip

Planning an experiment is an important skill to practise.

Make sure you clearly plan out what you're going to do and think about safety.

08 Cell **A** is a spherical animal cell with a radius of 5 μm. Cell **B** is also spherical. It has a radius of 20 μm.

08.1 Identify which statement about these two cells is true. Choose **one**. **[1 mark]**

Cell **A** has a smaller volume than cell **B**.

Cell **B** has a smaller surface area than cell **A**.

Cell **A** has a smaller surface area-to-volume ratio than cell **B**.

Cell **B** has a larger surface area-to-volume ratio than cell **A**.

08.2 Cell **A** has a surface-area-to-volume ratio of 0.6 : 1. It takes 5 ms for an amino acid to diffuse out of cell **A** into the bloodstream. Assuming that the rate of diffusion is proportional to the surface area-to-volume ratio of a cell, calculate the time taken for an identical amino acid to diffuse out of cell **B**. **[5 marks]**

 Exam Tip

There are lots of areas maths can be mixed with biology.

You may not have done this in class but don't let it worry you, just use your maths skills.

09 Plant roots absorb water from the soil by osmosis.

09.1 Define the term osmosis. **[1 mark]**

09.2 Once inside the root, water continues to move from cell to cell as it moves towards the xylem vessel. **Figure 6** shows three cells within the root. Each cell contains a different concentration of salt.

Figure 6

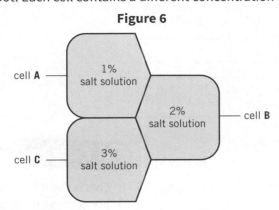

cell **A** —— 1% salt solution

2% salt solution —— cell **B**

cell **C** —— 3% salt solution

Water can move from cell to cell in any direction. Identify which cell will gain water the fastest. Give a reason for your answer. **[2 marks]**

09.3 Xylem vessels transport water throughout the plant. Describe the structure of the xylem vessels. **[2 marks]**

10 The alveoli in the lungs are adapted for gas exchange. One adaptation is a large surface-area-to-volume ratio.

10.1 Explain how a large surface-area-to-volume ratio maximises gas exchange. **[2 marks]**

10.2 Explain **one** other way the lungs are adapted for gas exchange. **[2 marks]**

10.3 Alveoli can be modelled as spheres. The diameter of an alveolus is 300 µm.

The surface area of a sphere is calculated using the formula: surface area = $4\pi r^2$.

The volume of a sphere is calculated using the formula: volume = $\frac{4}{3}\pi r^3$.

Calculate the surface area-to-volume ratio of an alveolus. **[4 marks]**

11 In many restaurants, vegetables are prepared in advance for the evening's meals. To prevent them turning brown, chefs often leave the prepared vegetables in slightly salted water. A chef wanted to know the ideal concentration of salt water to store potatoes. The chef used the following method:

1 Cut the potato into pieces of equal volume.

2 Measure the mass of each potato piece.

3 Place each potato piece into a different concentration of salt solution.

4 Leave for two hours.

5 Remove each potato piece and blot dry.

6 Measure the new mass of each potato piece.

The chef's results are shown in **Table 3**.

Table 3

Concentration of saltwater solution in M	0.0	0.5	1.0	1.5	2.0
Starting mass in g	2.8	3.0	3.1	2.9	2.9
Mass after 2 hours in g	3.1	3.0	3.0	2.7	2.4

11.1 Identify the solution in which the potato gained the most mass. **[1 mark]**

11.2 Explain why this potato gained mass. **[2 marks]**

11.3 Suggest which concentration of salt solution the chef should store the potato in. Explain your answer. **[2 marks]**

! **Exam Tip**

This is the difference between the starting mass and mass after two hours, not just the highest mass after two hours.

12 Substances such as water and ions need to move in and out of cells.

12.1 Draw **one** line between the process and the correct method of transport. **[2 marks]**

Process   Transport method

The movement of oxygen from the lungs into the bloodstream.		active transport
The movement of mineral ions from the soil into a plant root system.		osmosis
The movement of water into a plant cell.		diffusion

12.2 Complete the sentence using the correct bolded words. **[2 marks]**

Cells which carry out active transport contain **many / few** mitochondria so that there will be sufficient **chemicals / energy**.

12.3 Explain why active transport is required to move glucose from the small intestine into the bloodstream. **[3 marks]**

13   After eating, the body needs to absorb as much glucose from the digested food as possible. It is absorbed into the bloodstream via the villi cells in the small intestine. **Figure 7** shows the cell membrane of a villus cell in the small intestine.

Figure 7

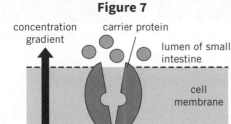
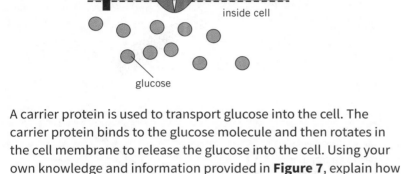
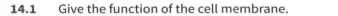

> **(!) Exam Tip**
>
> Lots of information is given to you in the diagram. Use this information to help you write your answer.

A carrier protein is used to transport glucose into the cell. The carrier protein binds to the glucose molecule and then rotates in the cell membrane to release the glucose into the cell. Using your own knowledge and information provided in **Figure 7**, explain how and why glucose is moved into the bloodstream by active transport and diffusion.   **[6 marks]**

14   Plant and animal cells share a number of common features.

14.1 Give the function of the cell membrane.   **[1 mark]**

14.2 Name **two** other structures found in both plant and animal cells.
                                                          [2 marks]

14.3 Chloroplasts are a sub-cellular structure found only in plant cells. A student uses a light microscope to observe cells from an onion bulb. Explain why the student is unable to view any chloroplasts.   **[3 marks]**

14.4 Plant cells also contain vacuoles. Explain how the vacuole helps plants to remain upright.   **[4 marks]**

B3 Cell division

Chromosomes

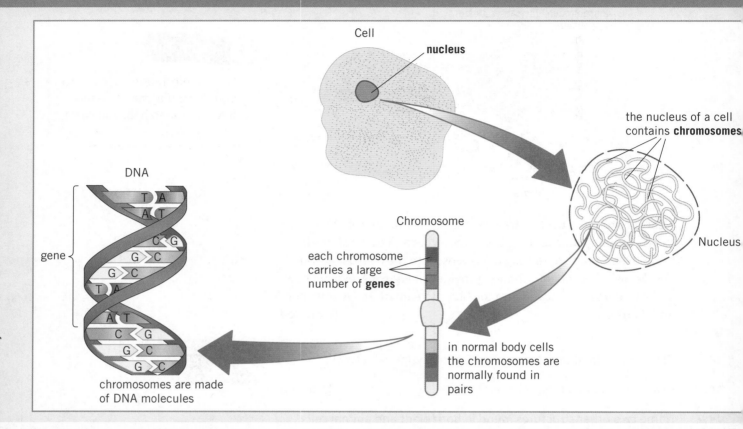

Cell

nucleus

the nucleus of a cell contains **chromosomes**

Nucleus

DNA

gene {

T A
A T
C G
G C
G C
T A
A T
C G
G C
G C

chromosomes are made of DNA molecules

Chromosome

each chromosome carries a large number of **genes**

in normal body cells the chromosomes are normally found in pairs

The cell cycle

Body cells divide to form two identical **daughter cells** by going through a series of stages known as the **cell cycle**.

Cell division by **mitosis** is important for the growth and repair of cells, for example, the replacement of skin cells. Mitosis is also used for asexual reproduction.

There are *three* main stages in the cell cycle:

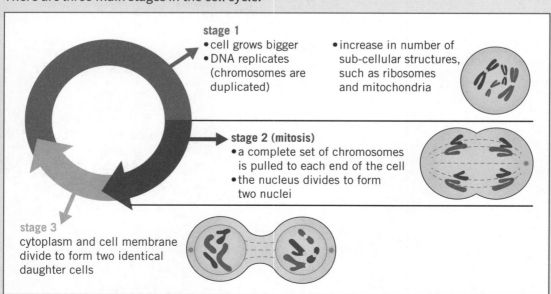

stage 1
- cell grows bigger
- DNA replicates (chromosomes are duplicated)
- increase in number of sub-cellular structures, such as ribosomes and mitochondria

stage 2 (mitosis)
- a complete set of chromosomes is pulled to each end of the cell
- the nucleus divides to form two nuclei

stage 3
cytoplasm and cell membrane divide to form two identical daughter cells

 Revision tip

Mitosis has a 't' in it, so that should help you remember that it makes *two* daughter cells.

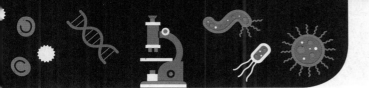

Stem cells in medicine

A stem cell is an undifferentiated cell that can develop into one or more types of specialised cell.

There are two types of stem cell in mammals: **adult stem cells** and **embryonic stem cells**.

Stem cells can be **cloned** to produce large numbers of identical cells.

Type of stem cell	Where are they found?	What can they differentiate into?	Advantages	Disadvantages
adult stem cells	specific parts of the body in adults and children – for example, bone marrow	can only differentiate to form certain types of cells – for example, stem cells in bone marrow can only differentiate into types of blood cell	• fewer ethical issues – adults can consent to have their stem cells removed and used • an established technique for treating diseases such as leukaemia • relatively safe to use as a treatment and donors recover quickly	• requires a donor, potentially meaning a long wait time to find someone suitable • can only differentiate into certain types of specialised cells, so can be used to treat fewer diseases
embryonic stem cells	early human embryos (often taken from spare embryos from fertility clinics)	can differentiate into any type of specialised cell in the body – for example, a nerve cell or a muscle cell	• can treat a wide range of diseases as can form any specialised cell • may be possible to grow whole replacement organs • usually no donor needed as they are obtained from spare embryos from fertility clinics	• ethical issues as the embryo is destroyed and each embryo is a potential human life • risk of transferring viral infections to the patient • newer treatment so relatively under researched – not yet clear if they can cure as many diseases as thought
plant meristem	meristem regions in the roots and shoots of plants	can differentiate into all cell types – they can be used to create clones of whole plants	• rare species of plants can be cloned to prevent extinction • plants with desirable traits, such as disease resistance, can be cloned to produce large numbers of identical plants • fast and low-cost production of large numbers of plants	• cloned plants are genetically identical, so a whole crop is at risk of being destroyed by a single disease or genetic defect

Therapeutic cloning

In **therapeutic cloning**

- cells from a patient's body are used to create a cloned early embryo of themselves
- stem cells from this embryo can be used for medical treatments and growing new organs
- these stem cells have the same genes as the patient, so are less likely to be rejected when transplanted.

 Key terms

Make sure you can write a definition for these key terms.

adult stem cell   cell cycle   chromosome

clone   daughter cells   embryonic stem cell

gene   meristem   mitosis

nucleus   therapeutic cloning

B3 Knowledge   25

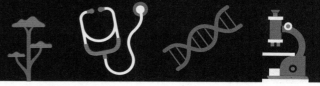
Learn the answers to the questions below then cover the answers column with a piece of paper and write as many as you can. Check and repeat.

B3 questions	Answers
1 What is a stem cell?	undifferentiated cell that can differentiate into one or more specialised cell types
2 What are adult stem cells?	stem cells from adults that can only differentiate into certain specialised cells
3 Where are adult stem cells found?	bone marrow
4 What are embryonic stem cells?	stem cells from embryos that can differentiate into any specialised cell
5 Where are embryonic stem cells found?	early human embryos (usually from spare embryos from fertility clinics)
6 What is therapeutic cloning?	patient's cells are used to create an early embryo clone of themselves – stem cells from the embryo can then be used to treat the patient's medical conditions
7 Give one advantage of using therapeutic cloning.	stem cells from the embryo are not rejected when transplanted because they have the same genes as the patient
8 Give one advantage of using adult stem cells.	fewer ethical issues as obtained from adults who can consent to their use
9 Give two disadvantages of using adult stem cells.	• can take a long time for a suitable donor to be found • can only differentiate into some specialised cell types, so treat fewer diseases
10 Give two advantages of using embryonic stem cells.	• can differentiate into any specialised cell, so can be used to treat many diseases • easier to obtain as they are found in spare embryos from fertility clinics
11 Give two disadvantages of using embryonic stem cells.	• ethical issues surrounding their use, as every embryo is a potential life • potential risks involved with treatments, such as transfer of viral infections
12 What are plant meristems?	area where rapid cell division occurs in the tips of roots and shoots
13 Give two advantages of using plant meristems to clone plants.	• rare species can be cloned to protect them from extinction • plants with special features (e.g., disease resistance) can be cloned to produce many copies
14 Give one disadvantage of using plant meristems to clone plants.	no genetic variation, so, for example, an entire cloned crop could be destroyed by a disease
15 What is cell division by mitosis?	body cells divide to form two identical daughter cells
16 What is the purpose of mitosis?	growth and repair of cells, and asexual reproduction
17 What happens during the first stage of the cell cycle?	cell grows bigger; chromosomes duplicate; number of sub-cellular structures (e.g., ribosomes and mitochondria) increases
18 What happens during mitosis?	one set of chromosomes is pulled to each end of the cell and the nucleus divides
19 What happens during the third stage of the cell cycle?	the cytoplasm and cell membrane divide, forming two identical daughter cells

Put paper here

B3

Now go back and use the questions below to check your knowledge from previous chapters.

Previous questions

Answers

Previous questions		Answers
Where is DNA found in animal and plant cells?	Put paper here	in the nucleus
What are two types of eukaryotic cell?		animal and plant
What is the function of the cell membrane?	Put paper here	controls movement of substances into and out of the cell
What is the function of mitochondria?		site of respiration to release energy for the cell
What is the function of chloroplasts?	Put paper here	contain chlorophyll to absorb light energy for photosynthesis
What is the function of ribosomes?	Put paper here	enable production of proteins (protein synthesis)
What is the function of the cell wall?		strengthens and supports the cell
What is diffusion?	Put paper here	net movement of particles from an area of high concentration to an area of low concentration along a concentration gradient – this is a passive process (does not require energy from respiration)
What is osmosis?	Put paper here	diffusion of water from a dilute solution to a concentrated solution through a partially permeable membrane

Maths Skills

Practise your maths skills using the worked example and practice questions below.

Converting Units

The size of a cell or organelle is most often shown in millimetres (mm), micrometres (μm), or nanometres (nm). You may be asked to convert between mm, μm and nm. If you are converting from a smaller unit to a larger unit, your number should get smaller. If you are converting a larger unit to a smaller unit, the number should get bigger.

Worked Example

- to convert mm to μm: multiply the mm reading by 1000
- to convert μm to nm: multiply the μm reading by 1000
- to convert nm into μm: divide the reading by 1000
- to convert μm into mm: divide the reading by 1000

Cell	Size in mm ← ÷1000	Size in μm ← ×1000	Size in nm
red blood cell	0.007	7	7000
leaf cell	0.06	60	60 000
egg cell	0.1	100	10

Practice

Convert the following cell and organelle sizes to complete the table.

Cell	Size in mm	Size in μm	Size in nm
ant	3		
human hair		100	
palisade leaf cell		70	
plant cell ribosome			20
HIV virus			100
egg cell mitochondria	0.002		

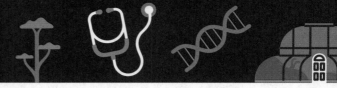
01   **Figure 1** shows some plant cells undergoing mitosis.

Figure 1

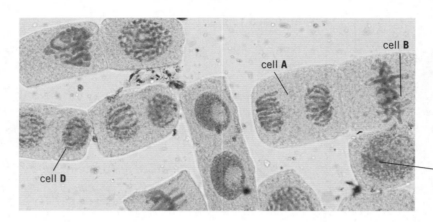

cell **B**

cell **A**

cell **D**

cell **C**

! Exam Tip

In mitosis you get two identical daughter cells.

01.1   Describe what is happening in cell **C**.   **[3 marks]**

_____

_____

_____

_____

! Exam Tip

Think about what needs to happen to the DNA before it can divide.

01.2   Identify which sequence of cells from **Figure 1** best represents the process of mitosis.   **[1 mark]**

Tick **one** box.

cell **D** → cell **A** → cell **B** → cell **C**   ☐

cell **C** → cell **B** → cell **A** → cell **D**   ☑

cell **A** → cell **B** → cell **C** → cell **D**   ☐

cell **C** → cell **D** → cell **B** → cell **A**   ☐

01.3   Cells **A**–**D** do not show the final stage of mitosis.

Describe what would happen at the next stage in this process.

[2 marks]

_____

_____

_____

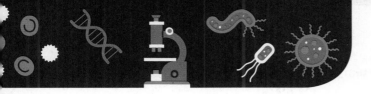 
01.4 When looking at cells under a microscope, the length of different stages of the cell cycle can be estimated using the formula:

$$\text{length of stage} = \frac{\text{number of cells at that stage} \times \text{total length of time in the cell cycle}}{\text{total number of cells}}$$

The average time taken for the plant cells in **Figure 1** to complete the cell cycle is 24 hours.

One stage in the mitosis cycle is called metaphase; this is where chromosomes line up at the centre of the cell.

Using the information in **Figure 1**, calculate the time taken for the metaphase stage. **[3 marks]**

> ! **Exam Tip**
>
> You may not have seen this equation before, but don't worry! You need to get used to using new and unfamiliar equations so you're ready for the exam.
>
> Just plug the numbers in and away you go!

_____ hours

02 People with Type 1 diabetes do not produce enough insulin. This is because the insulin-producing cells in the pancreas are destroyed by the body's immune system.

Patients with this form of diabetes have to inject themselves regularly with insulin.

Scientists hope that stem cells could be used to treat this condition one day.

02.1 Describe what is meant by a stem cell. **[1 mark]**

_____

_____

02.2 Suggest the role that stem cells could play in a diabetic person's body. **[1 mark]**

> ! **Exam Tip**
>
> Think about the cells that don't work properly in a diabetic.

_____

_____

02.3 A group of scientists carried out a study into the use of adult stem cells to treat Type 1 diabetes.

Describe the main difference between these stem cells and an embryonic stem cell. **[2 marks]**

_____

_____

_____

02.4 Suggest **one** reason why it is preferential to use stem cells from the actual patient instead of using cells from a donor. **[1 mark]**

_____

_____

02.5 The study used 23 patients. The patients taking part in the trial were tracked over a 30-month period. At the end of the investigation, 12 patients did not have to inject themselves with insulin anymore.

Calculate the percentage of patients for which the treatment was successful. **[1 mark]**

! Exam Tip

A common mistake when working out percentages is forgetting to multiple the answer by 100 at the end.

_____ %

02.6 Suggest and explain whether this technique is a successful treatment for Type 1 diabetes. **[1 mark]**

_____

_____

_____

03 All cells in the human body contain genetic information.

03.1 Describe how the genetic material is organised in the nucleus of a human cell. **[3 marks]**

03.2 As a baby grows, its cells change in a number of ways. Explain why mitosis and cell differentiation are important in the growth and development of a baby from a fertilised egg. **[4 marks]**

! Exam Tip

This is a four mark question, so try to write two points for each section.

03.3 Describe the main steps in mitosis. **[4 marks]**

04 Scientists hope that in the future it will be possible to use stem cells to help treat patients with a number of conditions, such as diabetes.

04.1 Explain why stem cells may be able to offer treatments for conditions such as diabetes that currently have no cure. **[3 marks]**

04.2 Name where in the human body stem cells can be found that can differentiate into different types of blood cell. **[1 mark]**

04.3 There are mixed opinions about the potential use of embryonic stem cells for the treatment of human diseases. Many people feel there are good reasons for carrying out this research, but others are opposed to these studies. Evaluate the ethical arguments surrounding the use of embryonic stem cells in medical research. **[6 marks]**

! Exam Tip

For an evaluate question, you need four key points:
1 the good things
2 the bad things
3 your opinion
4 the because (why you have that opinion).

If you don't include all of these, you won't get top marks.

05 Stem cells are found in both plants and animals.

05.1 Name the area in a plant where stem cells are located. **[1 mark]**

05.2 **Figure 2** shows a diagram of a root tip of a plant. Identify the letter (**A–D**) in **Figure 2** indicating where stem cells are located in the root tip. **[1 mark]**

Figure 2

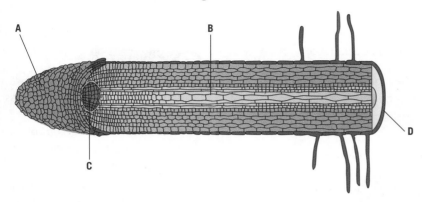

05.3 Identify the sequence that correctly describes the steps plant stem cells go through to produce a root hair cell. Choose **one** answer.

[1 mark]

A differentiation → DNA replication → elongation → mitosis

B elongation → differentiation → DNA replication → mitosis

C DNA replication → elongation → mitosis → differentiation

D DNA replication → mitosis → elongation → differentiation ✓

! Exam Tip

Work through these slowly, crossing off any you know are incorrect, and hopefully that will only leave one answer!

05.4 Tick **one** box in each row to identify the differences between plant and animal stem cell differentiation. **[2 marks]**

! Exam Tip

There is a clue in the question: "Tick one box in each row to…" means it's not the same answer for both!

	Animal stem cells	Plant stem cells
differentiation occurs at a very early stage	✓	
differentiation occurs throughout life		✓
differentiations produced are permanent	✓	
differentiation can be reversed or changed		✓

06.1 A scientist takes a cutting from a plant. Select the statement that best describes why the meristem in the cutting allows a new plant to grow. Choose **one** answer. **[1 mark]**

Meristems contain differentiated cells.

Meristems contain undifferentiated cells. ✓

Meristems are where new roots form.

Meristems respond to light.

06.2 Explan why this technique is a form of cloning. **[1 mark]**

06.3 Suggest **two** advantages of using this technique to produce new plants of this species, as opposed to letting the plant reproduce naturally. **[2 marks]**

07 When neurones in the brain stop producing dopamine, a person can develop Parkinson's disease. This is a disease of the nervous system. One of the symptoms experienced by sufferers is tremors (shaking).

07.1 Explain **two** ways in which nerve cells are specialised. **[4 marks]**

07.2 Dopamine is a chemical that allows neurones to communicate with each other. Suggest and explain why a person suffering from Parkinson's disease experiences tremors. **[2 marks]**

07.3 Therapeutic cloning is a type of stem cell research where scientists are trying to produce an early embryo clone from cells taken from an adult human. The stem cells from the patient's cloned embryo can then potentially be used to produce stem cells to treat a medical condition that the person has. This process is summarised in **Figure 3**.

Using your own knowledge and information in the question, suggest and explain how stem cell treatment could be used in the future to treat Parkinson's disease. **[3 marks]**

Figure 3

early human embryo

stem cells removed

stem cells cultured

stem cells made to differentiate into different tissues

spinal cord   heart   kidney   insulin-producing cells

organs or tissues transplanted into a patient to treat them

07.4 Discuss the advantages and disadvantages of using therapeutic cloning to treat Parkinson's disease. **[6 marks]**

08 Body cells divide in a regulated series of events called the cell cycle. The length of the cell cycle varies considerably between different types of cell and the stages of an organism's development.

08.1 Identify the cell with the shortest cell cycle. Choose **one** answer. **[1 mark]**

an adult nerve cell *A*   *B* a teenager's liver cell

a child's brain cell *C*   *D* an unborn foetal intestinal cell

> **Exam Tip**
>
> In the exam you'll come across topics you've never seen before and be expected to apply what you have learnt in your lessons.
>
> Lots of student are worried by this, but by using the questions in this book you're giving yourself a great head-start in this skill.
>
> Don't worry if you didn't cover dopamine or Parkinson's disease in class, this is a great chance to show off what you know about stem cell therapy.

> **Exam Tip**
>
> Lots of information is already in the question.
>
> Highlight the advantages one colour and the disadvantages another colour – this will help you easily pick out which bits you need to answer the question fully.

08.2 Some organs in the body contain cells that have a short cell cycle throughout a person's life. Identify **one** region in the body where this happens and give a reason for your answer. **[2 marks]**

! Exam Tip

Try to think of areas that frequently need new cells.

08.3 Describe the changes that occur at each stage of the cell cycle. **[4 marks]**

08.4 Suggest **one** reason why it is important that the chromosome number stays the same after mitosis. **[1 mark]**

! Exam Tip

Think about the process of cell division.

09 Scientists can now grow a number of different types of stem cells from embryonic stem cells. Many of the stem cells are taken from spare embryos created during fertility treatments, such as *in vitro* fertilisation (IVF).

09.1 Identify which type of microscope a scientist would use to check that an egg cell had been fertilised and developed into a healthy embryo. Give reasons for your answer. **[2 marks]**

! Exam Tip

Which type of microscope can view living cells?

09.2 Suggest **one** ethical concern some people may have with using embryonic stem cells. **[1 mark]**

09.3 A human egg cell is approximately 0.1 mm in diameter. A human sperm cell is approximately 2.5 μm. Calculate the difference in order of magnitude between a sperm cell and an egg cell. **[2 marks]**

10 **Figure 4** shows some cells taken from the root tips of an onion viewed under a microscope.

Figure 4

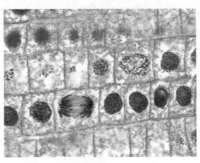

10.1 Describe how you can tell that these cells are undergoing mitosis. **[1 mark]**

! Exam Tip

Link your answer to what you can see in the image.

10.2 Explain how a microscope slide of root cells should be prepared for viewing. **[4 marks]**

10.3 The root cells were viewed using a 15 × eye piece lens and a 40 × objective lens. Calculate the total magnification used to view the root cells. **[1 mark]**

10.4 Root hair cells are an example of a specialised cell. Write down **one** way in which they are adapted for taking water into a plant. **[1 mark]**

! Exam Tip

Don't be tempted to use the equation for magnification here – this question is much simpler than that, even if you haven't come across this in class.

B4 Organisation in animals

There are five **levels of organisation** in living organisms:

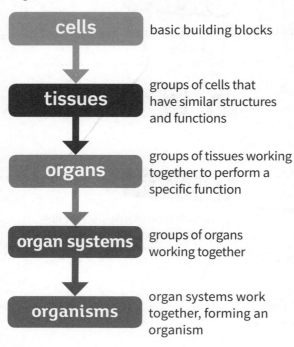

cells	basic building blocks
↓	
tissues	groups of cells that have similar structures and functions
↓	
organs	groups of tissues working together to perform a specific function
↓	
organ systems	groups of organs working together
↓	
organisms	organ systems work together, forming an organism

Digestive system

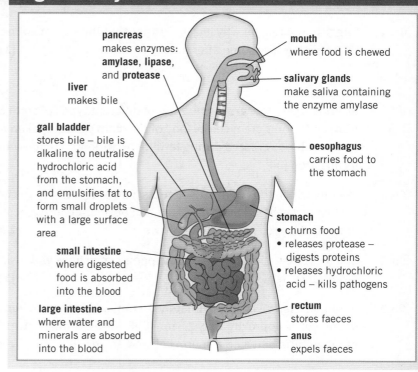

pancreas makes enzymes: **amylase**, **lipase**, and **protease**

liver makes bile

gall bladder stores bile – bile is alkaline to neutralise hydrochloric acid from the stomach, and emulsifies fat to form small droplets with a large surface area

small intestine where digested food is absorbed into the blood

large intestine where water and minerals are absorbed into the blood

mouth where food is chewed

salivary glands make saliva containing the enzyme amylase

oesophagus carries food to the stomach

stomach
- churns food
- releases protease – digests proteins
- releases hydrochloric acid – kills pathogens

rectum stores faeces

anus expels faeces

Lungs

When breathing in, air moves
1 into the body through the mouth and nose
2 down the trachea
3 into the **bronchi**
4 through the **bronchioles**
5 into the **alveoli** (air sacs).

Oxygen then diffuses into the blood in the network of **capillaries** over the surface of the alveoli.

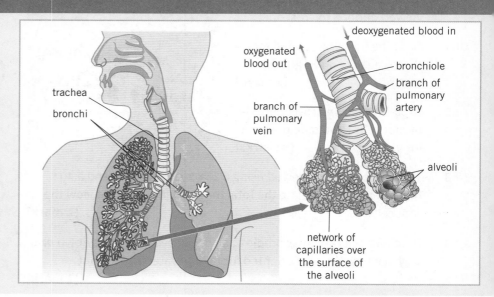

trachea

bronchi

oxygenated blood out

deoxygenated blood in

bronchiole

branch of pulmonary artery

branch of pulmonary vein

alveoli

network of capillaries over the surface of the alveoli

The circulatory system

blood is a tissue made up of four main components
- red blood cells – bind to oxygen and transport it around the body
- plasma – transports substances and blood cells around the body
- platelets – form blood clots to create barriers to infections
- white blood cells – part of the immune system to defend the body against pathogens

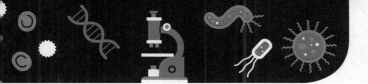

B4

Blood vessels

The structure of each blood vessel relates to its functions.

Vessel	Function	Structure	Diagram
artery	carries blood *away from* the heart (high pressure)	• thick, muscular, and elastic walls • the walls can stretch and withstand high pressure • small lumen	thick wall, small lumen, thick layer of muscle and elastic fibres
vein	carries blood *to* the heart (low pressure)	• have valves to stop blood flowing the wrong way • thin walls • large lumen	relatively thin wall, large lumen, often has valves
capillary	• carries blood to tissues and cells • connects arteries and veins	• one cell thick – short diffusion distance for substances to move between the blood and tissues (e.g., oxygen into cells and carbon dioxide out) • very narrow lumen	wall one cell thick, tiny vessel with narrow lumen

The heart

The heart is the organ that pumps blood around your body. It is made from **cardiac** muscle tissue, which is supplied with oxygen by the **coronary artery**.

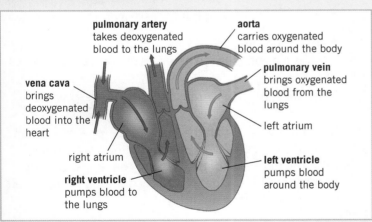

pulmonary artery takes deoxygenated blood to the lungs

aorta carries oxygenated blood around the body

pulmonary vein brings oxygenated blood from the lungs

left atrium

vena cava brings deoxygenated blood into the heart

right atrium

left ventricle pumps blood around the body

right ventricle pumps blood to the lungs

Heart rate is controlled by a group of cells in the right atrium that generate electrical impulses, acting as a pacemaker.

Artificial pacemakers can be used to control irregular heartbeats.

Double circulatory system

The human circulatory system is described as a **double circulatory system** because blood passes through the heart twice for every circuit around the body:

• the right ventricle pumps blood to the lungs where gas exchange takes place
• the left ventricle pumps blood around the rest of the body.

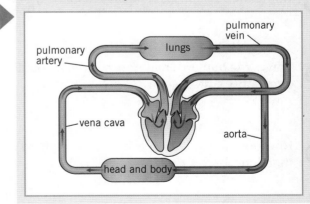

pulmonary vein

lungs

pulmonary artery

vena cava

aorta

head and body

 **Key terms**

Make sure you can write a definition for these key terms.

alveoli   amylase   aorta   artery   atrium   bronchi   bronchiole   capillary   cardiac
coronary   double circulatory system   lipase   organ   organ system   plasma   platelet
protease   pulmonary   tissue   vein   vena cava   ventricle

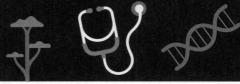
Learn the answers to the questions below, then cover the answers column with a piece of paper and write as many as you can. Check and repeat.

B4 questions	Answers
1 Name the five levels of organisation.	cells → tissues → organs → organ systems → organisms
2 What is a tissue?	group of cells with similar structures and functions
3 What is an organ?	group of tissues working together to perform a specific function
4 What is the function of the liver in digestion?	produces bile, which neutralises hydrochloric acid from the stomach and emulsifies fat to form small droplets with a large surface area
5 What is the function of saliva in digestion?	lubrication to help swallowing – contains amylase to break down starch
6 Name three enzymes produced in the pancreas.	amylase, protease, lipase
7 Name the four main components of blood.	red blood cells, white blood cells, plasma, platelets
8 What is the function of platelets?	form blood clots – prevent the loss of blood and stop wounds becoming infected
9 Describe three adaptations of a red blood cell.	• bi-concave disc shape – large surface area-to-volume ratio for diffusion of oxygen • contains haemoglobin – binds to oxygen • no nucleus – more space for oxygen
10 How do white blood cells protect the body?	• engulf pathogens • produce antitoxins to neutralise toxins, or antibodies
11 Name the substances transported in the blood plasma.	hormones, proteins, urea, carbon dioxide, glucose
12 Why is the human circulatory system a double circulatory system?	blood passes through the heart twice for every circuit around the body – deoxygenated blood is pumped from the right side of the heart to the lungs, and the oxygenated blood that returns is pumped from the left side of the heart to the body
13 How does the structure of an artery relate to its function?	carries blood away from the heart under high pressure – has a small lumen and thick, elasticated walls that can stretch
14 How does the structure of a vein relate to its function?	carries blood back to the heart at low pressure – doesn't need thick, elasticated walls, but has valves to prevent blood flowing the wrong way
15 How does the structure of a capillary relate to its function?	carries blood to cells and tissues – has a one-cell-thick wall to provide a short diffusion distance
16 List the structures air passes through when breathing in.	mouth/nose → trachea → bronchi → bronchioles → alveoli

Put paper here

B4

Now go back and use the questions below to check your knowledge from previous chapters.

Previous questions

Answers

What is the purpose of active transport in the small intestine?	Put paper here	sugars can be absorbed when the concentration of sugar in the small intestine is lower than the concentration of sugar in the blood
What is therapeutic cloning?		patient's cells are used to create an early embryo clone of themselves – stem cells from the embryo can then be used to treat the patient's medical conditions
What is a stem cell?	Put paper here	undifferentiated cell that can differentiate into one or more specialised cell types
Give one disadvantage of using plant meristems to clone plants.		no genetic variation, so, for example, an entire cloned crop could be destroyed by a disease
What is active transport?		movement of particles against a concentration gradient – from a dilute solution to a more concentrated solution – using energy from respiration

 Required Practical Skills

Practise answering questions on the required practicals using the example below. You need to be able to apply your skills and knowledge to other practicals too.

Food tests	Worked Example	Practice
There are different ways to test for four different compounds found in food: • ethanol test for lipids (fats) – colour change from colourless to cloudy if present • Benedict's test for sugars – colour change from blue to red if present • iodine test for starch (carbohydrates) – colour change from brown to blue-black if present • Biuret reagent test for protein – colour change from blue to purple if present. You need to be able to identify and describe the correct method, and results, for each test.	A student wanted to test a sample for the presence of protein using Biuret reagent. Write a risk assessment for this activity. **Answer:** Write down general safety practices in labs: • wear goggle to protect your eyes • wash hands at the end of the practical • clear up any spills quickly • do not eat any of the food Write down what things could hurt you in the practical, and how they could hurt you: • Biuret reagent – irritant • glass – can break • pipette – can poke you in the eyes Write down how you can prevent these hurting you: • wash hands after touching Biuret reagent, do not eat in the lab, and if ingested or it gets into the eyes inform the teacher immediately • if glass is broken inform a teacher immediately • point pipettes downwards	1 A student picked up solution A and added it to a sample of food. Solution A was blue and turned purple after adding to the food. Name solution A, and identify the food present in the sample. 2 Benedict's test for sugar requires the solution to be heated. One way to do this is heating the test tube in a beaker of water using a Bunsen burner. Give an alternative method of heating the solution. 3 When testing a sample for protein in a test tube, a student found that the top of the sample tested positive whereas the bottom did not. Give the reason for this result.

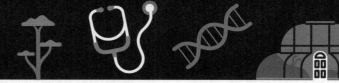
01   The events that occur during one breath – one inhalation and one exhalation – are known as one respiratory cycle.

Figure 1 shows change in the volume of the lungs in one respiratory cycle. The data were taken when the person was resting.

Figure 1

> **! Exam Tip**
>
> Draw lines on the graph to help you work it out!

01.1   Use **Figure 1** to determine the volume of air taken in when the person inhales.   **[1 mark]**

_____

_____ dm³

01.2   The person's total lung volume after inhalation was 6.00 dm³. Calculate their total lung volume after exhalation.   **[2 marks]**

_____

_____

_____ dm³

01.3   Calculate how many respiratory cycles will take place in one minute. Give your answer to **two** significant figures.   **[3 marks]**

_____

_____

_____

_____ per minute

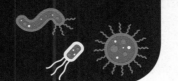

01.4 Explain how the structures in the chest cavity cause the changes in lung volume shown between 0 s and 1 s. **[4 marks]**

_____

_____

_____

_____

_____

Exam Tip

An 'explain' question, wants to know why things are happening.

01.5 A doctor measured another person's resting respiratory cycle. This person had 25 respiratory cycles per minute.

Suggest and explain **one** possible cause of this difference. **[2 marks]**

_____

_____

_____

Exam Tip

'Suggest and explain' means you need to say *what* you think will happen and *why*.

02 A student carried out a number of food tests on an unknown sample. Their results are shown in **Table 1**.

Table 1

Reagent used	Result
iodine	yellow–orange
Benedict's solution	blue
Biuret reagent	purple
ethanol	cloudy white layer formed

02.1 Suggest and explain **one** safety precaution that the student should have taken when using the Biuret reagent. **[2 marks]**

_____

_____

_____

Exam Tip

The question has asked for a specific safety precaution when using Biuret reagent, so a general safety measure isn't going to get the marks!

02.2 Identify which of the following statements is a correct description of the student's findings.

Tick **one** box. **[1 mark]**

The food sample contains starch, protein, and fat. ☐

The food sample contains starch and sugar. ☐

The food sample contains fat and protein. ☐

The food sample contains fat and sugar. ☐

02.3 Identify which foodstuff is most likely to be the food sample the student tested.

Tick **one** box. **[1 mark]**

a chocolate bar ☐

a meat burger ☐

spaghetti ☐

a carrot ☐

02.4 Many people with diabetes have to follow a strict diet to control their blood glucose levels.

Explain why using Benedict's solution to test foods for glucose may not be helpful to a diabetic. **[2 marks]**

_____

_____

_____

! Exam Tip

Think about which food diabetics need to control their intake of.

03 Gluten is a form of protein found in some grains, for example, wheat.

03.1 Describe the structure of a protein. **[1 mark]**

03.2 Coeliac disease is a disease of the digestive system. It damages the lining of the small intestine when foods that contain gluten are eaten, resulting in a patient having a reduced number of villi. This causes a number of symptoms such as abdominal bloating and pain. A healthy person has on average 25 to 30 villi per μm^2. Calculate the density of the villi in the small intestine of a coeliac patient who has 50 000 villi in 7200 μm^2 of small intestine. **[2 marks]**

! Exam Tip

For this question coeliac disease is used as an example. You may not have covered this in class, but this is getting you used to applying what you know to new situations for the exam.

03.3 **Figure 2** compares a section of the small intestine of a person with coeliac disease with a person who does not have coeliac disease.

Figure 2

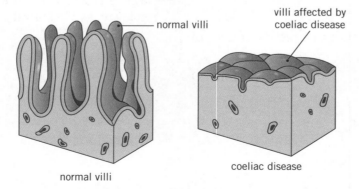

normal villi

villi affected by coeliac disease

normal villi

coeliac disease

! Exam Tip

There is a clear difference in the picture. Think about how this difference may relate to the function of the digestive system.

Use the information in the question and your own knowledge to suggest why a child with coeliac disease may not grow as tall as their peers. **[4 marks]**

04 **Figure 3** shows some organs from the digestive system.

04.1 Identify organs **A** and **C** from **Figure 3**. [2 marks]

04.2 Identify the organ from **Figure 3** that is responsible for absorbing water from undigested food. [1 mark]

04.3 The stomach is made up of a number of tissues. Draw **one** line from each type of stomach tissue to its function. [3 marks]

Figure 3

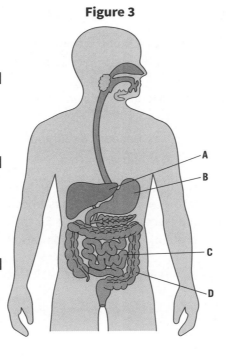

Stomach tissue		Function
		churns the food and digestive juices of the stomach together
muscular tissue		
		covers the inside and outside of the stomach
glandular tissue		
		sends impulses to other areas of the body
epithelial tissue		
		produces the digestive juices

! Exam Tip

Be careful – there is one spare function!

! Exam Tip

Your answer must relate to fat digestion, not to any other functions!

04.4 Explain how the pancreas and the gall bladder work together to increase the efficiency of fat digestion. [6 marks]

05 **Figure 4** represents cross-sectional areas through the three main types of blood vessel.

Figure 4

A   B   C

05.1 Identify which blood vessel in **Figure 4** represents an artery. [1 mark]

05.2 Explain **one** way arteries are adapted for their function. [2 marks]

! Exam Tip

Use **Figure 4** to help with this question!

05.3 Blood in the arteries is usually bright red because it is full of oxygen. Identify the artery where this is not true. [1 mark]

aorta   vena cava   pulmonary artery   coronary artery

05.4 Give a reason for your answer to **05.3**. [1 mark]

05.5 Describe **two** reasons why it is important that blood is transported to every cell in the body. [2 marks]

06 A student was provided with an unknown food sample and the following apparatus:

- test tubes (in a test-tube rack)
- water bath
- iodine
- Benedict's solution
- Biuret reagent
- ethanol

The food sample had been ground into a powder using a pestle and mortar. Explain how the student could test the food sample for the presence of starch, sugar, fats, and protein. [6 marks]

07 **Figure 5** shows a cross-section through the human heart.

Figure 5

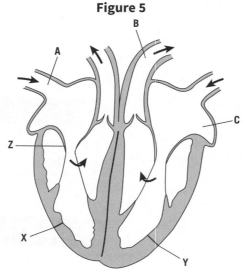

magnification: × 0.75

Exam Tips

This is a great practice question.

Remember, don't use all of your sample in one go, clearly lay out what observations would mean positive or negative results, and don't forget your safety precautions.

Exam Tip

The first thing you should do when you see a diagram of a heart is mark down your right (on the left-hand side) and left (on the right-hand side).

07.1 Identify which label is pointing to the left atrium. [1 mark]

07.2 Name the blood vessels labelled **A** and **B**. [2 marks]

07.3 Identify and describe the function of part **Z**. [2 marks]

07.4 Humans have a double circulatory system. Describe what this means. [2 marks]

08 **Table 2** shows the number of red blood cells present in people living at different altitudes above sea level.

Table 2

Height above sea level in m	Mean number of red blood cells in millions per mm³ of blood
0	4.9
1000	5.5
2000	6.2
3000	6.8
4000	7.2
5000	7.6

08.1 Explain how a red blood cell is adapted to perform its function. **[6 marks]**

08.2 Using **Table 2**, calculate the percentage change between the number of red blood cells present in a person living at 2000 m above sea level, and 4000 m above sea level. **[2 marks]**

08.3 As altitude increases, the amount of oxygen in the air decreases. Using information in **Table 2**, explain how differences in a person's blood composition enable them to live at different altitudes. **[3 marks]**

09 **Figure 6** shows a section of blood vessels in the upper arm.

Figure 6

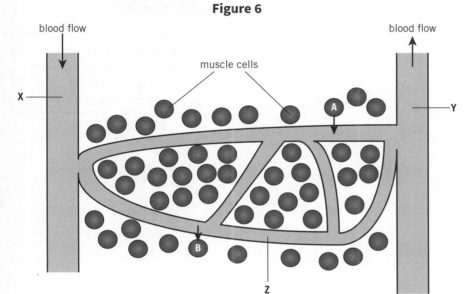

09.1 Name the blood vessel represented by label **Z**. **[1 mark]**

09.2 Describe **two** ways structure **Z** is adapted to maximise the rate of diffusion of carbon dioxide. **[2 marks]**

09.3 Identify which arrow shows the direction of transport of carbon dioxide. **[1 mark]**

10 **Figure 7** shows the changes in blood pressure of a person at rest. Their blood pressure was measured in an artery and a vein at the same time.

Figure 7

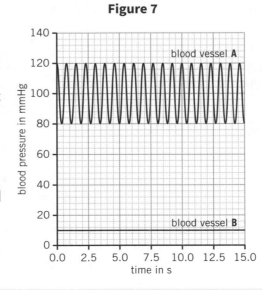

10.1 Identify which line on the graph represents a vein. Explain your answer. **[3 marks]**

10.2 Explain how a vein is adapted to transporting blood back to the heart. **[4 marks]**

10.3 Draw a line on **Figure 7** to suggest the pressure you would find in a capillary. **[1 mark]**

10.4 Calculate the person's heart rate. **[2 marks]**

11.1 Explain why a household plant will start to wilt if you don't water it for several days. **[4 marks]**

11.2 Plants also need a supply of minerals to remain healthy, for example, nitrates. Nitrates are taken in through the roots. **Figure 8** shows the change over time in the mass uptake of nitrate ions for root cells with and without access to oxygen.

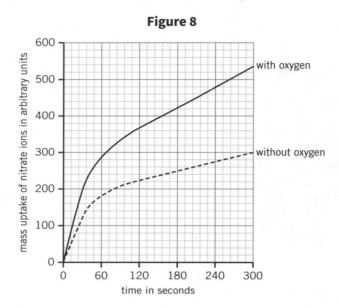

Figure 8

Describe the trends shown in the mass uptake of nitrate ions. **[4 marks]**

11.3 Calculate the percentage difference in the mass of nitrate uptake for roots with and without access to oxygen after 210 seconds. **[2 marks]**

11.4 Explain why the plant had taken up a greater mass of nitrate ions at 210 seconds when a supply of oxygen was present. **[4 marks]**

12 A student wanted to observe some of his own cells in the classroom. He was told to use skin cells from the back of his hand. He used a piece of clear sticky tape to remove some dead cells, which he placed on a microscope slide.

12.1 Describe how the student should use the microscope to observe the slide. **[3 marks]**

12.2 Draw a labelled diagram of the cell the student would expect to view through the microscope. **[4 marks]**

12.3 Name **one** additional structure the student would be able to see if he observed the skin cells using an electron microscope. Give the function of this structure. **[2 marks]**

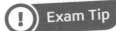 Exam Tip

Measuring the number of beats in 15 seconds then multiplying by four will give a more accurate result.

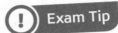 Exam Tip

Give the story of the lines, talk about both lines, use figures from the graph to show where any changes take place, and describe the shape of the graph.

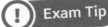 Exam Tip

This may be very similar to a practical you've done in class, but it is slightly changed so you can practise applying what you know to new situations.

 Exam Tip

Remember this is an animal cell, not a plant cell!

12.4 Suggest **one** reason why the student would not be told to observe their own blood cells in the classroom. **[1 mark]**

13 The best light microscopes produce magnifications of around ×2000. Transmission electron microscopes (TEMs) have much better resolution, producing magnifications of around ×2 000 000.

13.1 Define the term resolution. **[1 mark]**

13.2 The smallest object that can be viewed through a light microscope is 200 nm in size. Calculate the smallest object that can be viewed through a transmission electron microscope. **[3 marks]**

13.3 Evaluate and compare the use of light and electron microscopes to observe sperm cells. **[6 marks]**

> **! Exam Tip**
>
> 'Compare' means you need to give the things that are the same and things that are different.

14 **Figure 9** shows the typical lengths of some different structures.

Figure 9

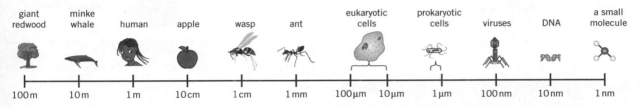

giant redwood | minke whale | human | apple | wasp | ant | eukaryotic cells | prokaryotic cells | viruses | DNA | a small molecule

100 m | 10 m | 1 m | 10 cm | 1 cm | 1 mm | 100 μm | 10 μm | 1 μm | 100 nm | 10 nm | 1 nm

14.1 Deduce the difference in order of magnitude between an ant and a human. **[1 mark]**

14.2 Explain why humans need lungs whereas ants do not. **[2 marks]**

14.3 The resolution of a light microscope is 800 nm. Identify the smallest structure from **Figure 9** that can be viewed using the light microscope. **[2 marks]**

14.4 An angstrom (Å) is a unit of length used to measure small distances. $1 Å = 1 \times 10^{-10}$ m.

Hydrogen atoms have a diameter of 1 Å. Calculate how many hydrogen atoms, placed end to end, would be the same width as a virus cell. **[3 marks]**

> **! Exam Tip**
>
> This may be a new unit to you, but don't let that distract you or put you off. Just treat it like you would any other calculation question.

B5 Enzymes

Enzymes

Enzymes are large proteins that **catalyse** (speed up) reactions. Enzymes are not changed in the reactions they catalyse.

Lock and key theory

This is a simple model of how enzymes work:

1 The enzyme's **active site** (where the reaction occurs) is a specific shape.
2 The enzyme (the lock) will only catalyse a specific reaction because the **substrate** (the key) fits into its active site.
3 At the active site, enzymes can break molecules down into smaller ones or bind small molecules together to form larger ones.
4 When the products have been released, the enzyme's active site can accept another substrate molecule.

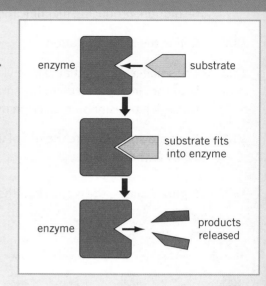

The effect of temperature on enzymes

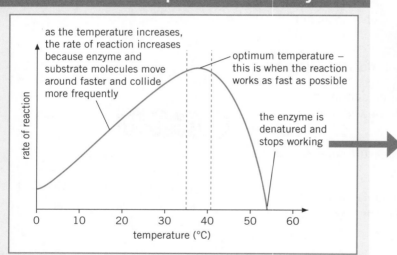

as the temperature increases, the rate of reaction increases because enzyme and substrate molecules move around faster and collide more frequently

optimum temperature – this is when the reaction works as fast as possible

the enzyme is denatured and stops working

🐾 Revision tip

This is one area where biology and chemistry overlap.

The first part of the graph can be explained by the collision theory you have learnt in your chemistry lessons.

Denaturation

At extremes of pH or at very high temperatures the shape of an enzyme's active site can change.

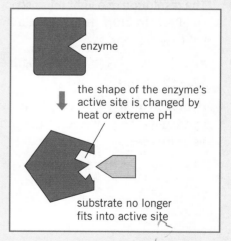

enzyme

the shape of the enzyme's active site is changed by heat or extreme pH

substrate no longer fits into active site

The substrate can no longer bind to the active site, so the enzyme cannot catalyse the reaction – the enzyme has been **denatured**.

🔑 Key terms

Make sure you can write a definition for these key terms.

active site   amylase   catalyse   denatured   enzyme

Digestive enzymes

Digestive enzymes convert food into small, soluble molecules that can then be absorbed into the bloodstream. For example, carbohydrases break down carbohydrates into simple sugars.

These products of digestion can be used to build new carbohydrates, lipids, and proteins.

Some of the glucose produced is used in respiration.

Enzyme	Sites of production	Reaction catalysed
amylase	salivary glands pancreas small intestine	starch → glucose
proteases	stomach pancreas small intestine	proteins → amino acids
lipases	pancreas small intestine	lipids → fatty acids and glycerol

The effect of pH on enzymes

Different enzymes have different optimum pH values.

This allows enzymes to be adapted to work well in environments with different pH values. For example, parts of the digestive system greatly differ in pH.

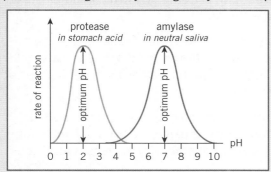

 **Revision tip**

When you're talking about enzymes, it's really important that you use the correct terms.

When the active site breaks down an enzyme becomes denatured – lots of students write that the enzyme has died, or has been killed.

This is incorrect and will lose you marks in the exam.

Revision tip

These graphs are very common exam questions.

Make sure you can draw them, recognise their shapes, and explain fully what is going on in each part of the graph.

lipase   optimum   protease   substrate

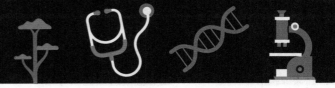
Learn the answers to the questions below then cover the answers column with a piece of paper and write as many as you can. Check and repeat.

	B5 questions		Answers
1	What are enzymes?	Put paper here	protein molecules that catalyse specific reactions in organisms
2	Why are enzymes described as specific?	Put paper here	each enzyme only catalyses a specific reaction, because the active site only fits together with certain substrates (like a lock and key)
3	Describe the function of amylase.	Put paper here	to break down starch into glucose
4	Where is amylase produced?	Put paper here	salivary glands, pancreas, and small intestine
5	Describe the function of proteases.	Put paper here	to break down proteins into amino acids
6	Where are proteases produced?	Put paper here	stomach, pancreas, and small intestine
7	Describe the function of lipases.	Put paper here	to break down lipids into fatty acids and glycerol
8	Where are lipases produced?	Put paper here	pancreas and small intestine
9	What are two factors that affect the rate of activity of an enzyme?	Put paper here	temperature and pH
10	What does denatured mean?	Put paper here	shape of an enzyme's active site is changed by high temperatures or an extreme pH, so it can no longer bind with the substrate
11	Describe the effect of temperature on enzyme activity.	Put paper here	as temperature increases, rate of reaction increases until it reaches the optimum for enzyme activity – above this temperature enzyme activity decreases and eventually stops
12	Describe the effect of pH on enzyme activity.	Put paper here	different enzymes have a different optimum pH at which their activity is greatest – a pH much lower or higher than this enzyme activity decreases and stops
13	Why do different digestive enzymes have different optimum pHs?	Put paper here	different parts of the digestive system have very different pHs – the stomach is strongly acidic, and the pH in the small intestine is close to neutral

Now go back and use the questions below to check your knowledge from previous chapters.

B5

Previous questions | Answers

What is the function of saliva in digestion?	lubrication to help swallowing; contains amylase to break down starch
Why is active transport needed in plant roots?	concentration of mineral ions in the soil is lower than inside the root hair cells – the mineral ions must move against the concentration gradient to enter the root hair cells
What happens during mitosis?	one set of chromosomes is pulled to each end of the cell and the nucleus divides
Where are embryonic stem cells found?	early human embryos (usually from spare embryos from fertility clinics)
How does the structure of an artery relate to its function?	carries blood away from the heart under high pressure – has a small lumen and thick, elasticated walls that can stretch
What is the function of a nerve cell?	carries electrical impulses around the body
What are plant meristems?	area where rapid cell division occurs in the tips of roots and shoots
Name the five levels of organisation.	cells → tissues → organs → organ systems → organisms

Put paper here

Required Practical Skills

Practise answering questions on the required practicals using the example below. You need to be able to apply your skills and knowledge to other practicals too.

Rate of enzyme reaction	Worked example	Practice
This practical tests your ability to accurately measure and record time, temperature, volume, and pH.	A class carried out an investigation into the effect that pH has on the ability of amylase to break down carbohydrates. They timed how long it took for the amylase to break down starch at different pH values between 5 and 11. Suggest the results the class would observe.	1 A student wanted to repeat the experiment on the following day to compare their results. Suggest why using the same enzyme solution on two different days would not give comparable results.
You will need to know how to find the rate of a reaction by using a continuous sampling technique to measure the time taken for an indicator to change colour.	**Answer:**	2 Suggest how the class might have timed how long it took for the amylase to break down the starch.
You will be familiar with measuring the effect of pH on the rate of reaction of amylase digesting starch, using iodine as an indicator. However, you need to be able to apply the methods of this practical to different enzymes and substrates!	Optimal pH of amylase is around 7, so the time taken to break down starch will be shortest at pH 7. At pH values lower than 7 it will take longer to break down the starch – it will take the longest time at pH 5, decreasing in time taken until pH 7. Above pH 7 it will take a longer time to break down the starch, and the amylase may stop breaking down the starch entirely at pH 11.	3 Give one variable the class must control for this experiment to be valid.

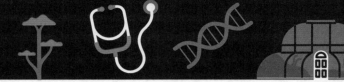

Exam-style questions

01 Lipase is an enzyme that breaks down lipids.

01.1 Name the products when a lipid is broken down. **[1 mark]**

_____

01.2 Name **one** organ in the body where lipase is made. **[1 mark]**

_____

> **! Exam Tip**
>
> Can you think of another name for a lipid that will point you towards the answer?

01.3 A group of students investigated the effect of temperature on the action of the enzyme lipase.

They used the following method in their investigation:

1   Add 10 cm³ of lipid solution to a test tube.

2   Add 2 cm³ of lipase solution to a second test tube.

3   Place both test tubes into a water bath set at 20 °C.

4   Leave in the water bath for five minutes.

5   Add the lipid solution to the lipase solution and mix.

6   Remove a sample of the mixture every five minutes and test for the presence of lipids. Continue until no lipid is detected.

7   Repeat the experiment every 5 °C between temperatures of 20 °C and 50 °C.

Name the independent variable in the students' investigation. **[1 mark]**

_____

01.4 Suggest why the lipase solution and lipid solution were left in the water bath for five minutes before mixing. **[1 mark]**

_____

_____

01.5 The students' results are shown in **Table 1**.

Table 1

Temperature in °C	Mean time taken until no lipid remained in min
20	20
25	15
30	10
35	5
40	10
45	20
50	lipid still present after 30 minutes of testing

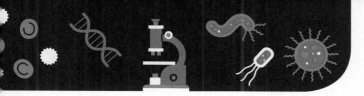

Describe the effect on the breakdown of the lipid when the temperature is increased between 20 °C and 35 °C. **[1 mark]**

_____

_____

! Exam Tip

You can make a quick sketch of the graph if you think it will help answer this question.

01.6   Explain the result that was observed at 50 °C. **[2 marks]**

_____

_____

_____

! Exam Tip

This question says 'explain' – it is asking _why_ that results happened not just _what_ happened.

02   **Figure 1** demonstrates the lock and key theory of enzyme action.

Figure 1

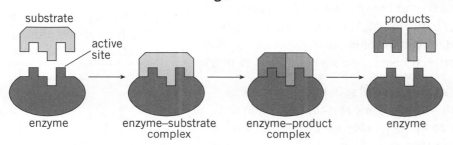

02.1   Using **Figure 1** and your own knowledge, explain what is meant by enzyme specificity. **[3 marks]**

_____

_____

_____

_____

! Exam Tip

There are lots of key words in the diagram – make sure you use them all in your answer!

02.2   Explain why you only need a small volume of an enzyme to catalyse a reaction. **[2 marks]**

_____

_____

02.3   Describe **one** example of an enzyme-controlled reaction where small molecules are joined together to form larger ones. **[1 mark]**

_____

_____

02.4 Measles is an infectious disease caused by a virus.
It causes sufferers to have a raised body temperature. Using your
knowledge of enzymes, suggest and explain **one** way in which this
may be damaging to the body and **one** way in which this may be
beneficial to the body. **[4 marks]**

_____

_____

_____

_____

_____

03 A group of students investigated the effect of pH on the action of
the enzyme amylase.

03.1 Name the substance that is broken down by amylase. **[1 mark]**

! Exam Tip

Go through the text with a
highlighter and pick out anything
that was kept the same.

03.2 The students placed starch solutions of known volume and
concentration in a water bath at 30 °C. They then added a buffer
solution, at one of five different pH values, to each starch solution.
Give **two** variables that the students controlled. **[2 marks]**

03.3 The students then took each sample of starch solution, one at
a time, and mixed it with a fixed volume and concentration of
amylase. They used the equipment in **Figure 2** to test for the
presence of starch every 30 seconds.

Figure 2

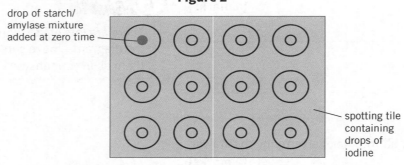

drop of starch/
amylase mixture
added at zero time

spotting tile
containing
drops of
iodine

Describe how you would monitor the reaction to identify when all
of the starch has been broken down. **[3 marks]**

03.4 The students' results are shown in **Table 2**.

Table 2

pH of buffer solution	Time taken for amylase to break down starch solution in s			
	Repeat 1	Repeat 2	Repeat 3	Mean
5	112	120	119	117
6	33	30	27	30
7	33	28	29	30
8	55	65	60	60
9	129	120	135	_128_

Calculate the mean time taken for the action of amylase
at pH 9. **[1 mark]**

03.5 Plot the students' mean results on **Figure 3**. **[3 marks]**

03.6 Use **Figure 3** to calculate the optimum pH for amylase to catalyse the breakdown of starch. **[1 mark]**

Figure 3

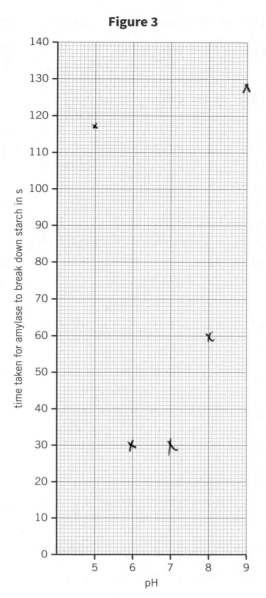

time taken for amylase to break down starch in s

pH

04 **Figure 4** shows how pH affects the activity of two different types of protease enzyme – enzyme **A** and enzyme **B**.

04.1 Name the substance that proteases break down into amino acids. **[1 mark]**

04.2 Describe the role of amino acids in the body. **[2 marks]**

Figure 4

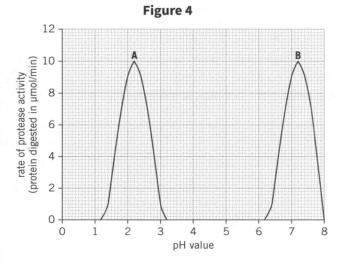

rate of protease activity (protein digested in μmol/min)

pH value

04.3 Use **Figure 4** to identify the optimum pH of enzyme **A**. **[1 mark]**

Exam Tip

Draw construction lines on your graph – this is your working out!

04.4 Suggest and explain where enzymes **A** and **B** are found in the body. **[4 marks]**

04.5 Explain the advantage of adding enzymes to biological washing powders. **[4 marks]**

04.6 Explain why many biological washing powders recommend not washing clothes on a 60 °C cycle. **[2 marks]**

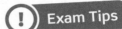
Exam Tip

Think about enzyme action at high temperatures.

05 A student was studying the effect of pH on the enzyme activity of an unknown carbohydrase. They were provided with the following apparatus:

- test tubes and rack
- spotting tiles
- 10 cm³ measuring cylinder
- 3 cm³ pipettes
- glass stirring rod
- stopwatch
- safety goggles
- starch solution
- carbohydrase solution
- iodine solution
- thermometer
- pH buffer solutions.

Explain how the student could investigate the effect of pH on the rate of reaction of the enzyme. **[6 marks]**

Exam Tips

Practice at planning experiments is if vital for exam success!

Plan a clear step-by-step method that could be followed by another person, stating volumes and equipment, and any safety precautions.

06 Biological washing powders contain enzymes. A scientist carried out an investigation to determine if a new type of protease enzyme should be included in washing powder.

06.1 Describe the function of proteases. **[1 mark]**

06.2 Protease function can be studied by looking at the time it takes to digest cooked egg white.

- The scientist placed a 2 cm³ piece of egg white into a test tube.
- They then added a fixed volume of the protease enzyme to the test tube and timed how long it took for the egg white to halve in length.
- The experiment was repeated at temperatures between 10 °C and 60 °C.
- A control was also set up using water instead of protease at each temperature. The egg white in the control samples remained undigested after two hours.

Name the equipment the scientist should have used to change the temperature. **[1 mark]**

06.3 **Figure 5** shows the scientist's results.

Figure 5

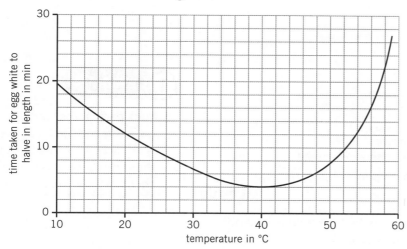

Identify the optimum temperature for protease activity. **[1 mark]**

06.4 Calculate the rate of reaction for the enzyme to break down the egg white at 20 °C. Give the unit of rate. **[3 marks]**

06.5 Using information in **Figure 5** and your own knowledge, suggest and explain **one** advantage and **one** disadvantage of using this enzyme in a biological washing powder. **[4 marks]**

07 Hydrogen peroxide, H_2O_2, is a by-product of cellular respiration and is made by all living cells. Hydrogen peroxide can be harmful and is normally removed as soon as it is produced in the cell. Cells make the enzyme catalase to remove hydrogen peroxide. Catalase catalyses the reaction:

hydrogen peroxide → water + oxygen

07.1 Explain why catalase is referred to as a catalyst. **[2 marks]**

07.2 A group of students investigated the action of catalase on hydrogen peroxide at different temperatures. Both catalase and hydrogen peroxide are at a fixed concentration and pH. They used the following method, as shown in **Figure 6**:

1 Add 1 cm³ hydrogen peroxide solution to a test tube.

2 Add 1 cm³ of catalase solution.

3 Foam containing oxygen bubbles will then be produced above the surface of the liquid.

Exam Tip

You may be surprised to see this question in biology, but we know the exam is going to be full of surprises! It's the same method we use in chemistry to find the gradient.

4   Measure the maximum height of the foam produced.

Figure 6

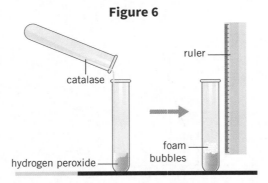

Describe what would happen if a lit splint was inserted into the foam. **[1 mark]**

Exam Tip

Look back at the start of the question and think about the gases that are going to be in those bubbles. This is a bit of chemistry knowledge!

07.3   The students controlled the volumes of hydrogen peroxide and catalase used. Name **one** other control variable. **[1 mark]**

07.4   The students repeated the experiment at a range of temperatures. Their results are shown in **Table 3**.

Table 3

Temperature in °C	Maximum foam height in cm			
	Repeat 1	Repeat 2	Repeat 3	Mean
10	1.7	1.4	1.4	1.5
20	3.2	3.1	2.7	3.0
30	1.8	5.2	4.8	
40	3.9	4.3	3.8	4.0
50	1.5	1.6	1.4	1.5
60	0.0	0.0	0.0	0.0

Explain why the students carried out the experiment three times at each temperature. **[1 mark]**

Exam Tip

Whenever you're asked to calculate a mean, always be on the look out for anomalous results.

07.5   Complete **Table 3** by calculating the mean result for 30 °C. **[1 mark]**

07.6   Explain the results at 60 °C. **[2 marks]**

Exam Tip

Why was there no foam?

07.7   The students concluded that the optimum temperature for catalase activity was between 30 °C and 40 °C. Explain how the students could improve their investigation to find a more accurate value for the optimum temperature for catalase activity. **[2 marks]**

08   Trypsin is an example of a protease enzyme.

08.1   Name the type of molecule broken down by trypsin. **[1 mark]**

08.2   Trypsin is produced in the pancreas and released into the small intestine. Identify the optimum pH for trypsin activity. Choose **one** answer. **[1 mark]**

pH 2   pH 4   pH 8   pH 9

Exam Tip

Key words are important for this question.

08.3 Trypsin is specific for catalysing one type of reaction. Using the lock and key theory, explain what is meant by enzyme specificity. **[3 marks]**

09 Cryophilic bacteria are a group of bacteria capable of growing and reproducing at low temperatures, ranging from −20 °C to +10 °C. They are found in permanently cold environments such as polar regions and the deep sea. They are able to survive because their enzymes are able to work at low temperatures.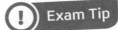

 **Exam Tip**

Don't worry if you've never heard of cryophilic bacteria before. This is just about applying the science you know to a new context.

09.1 On **Figure 7**, draw and label a line to represent the rate of reaction at different temperatures of an enzyme found in humans. **[2 marks]**

Exam Tip

Your line needs to be bell-shaped!

Figure 7

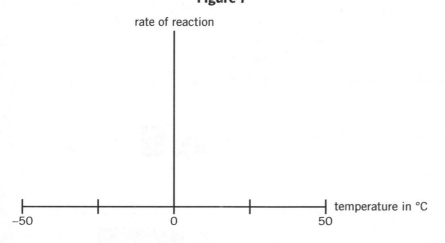

Exam Tip

This graph may look a bit confusing, but that's just because the y-axis is in the middle. Treat it like any other graph.

09.2 Draw and label a second line on **Figure 7** to represent the rate of reaction at different temperatures of an enzyme found in cryophilic bacteria. **[2 marks]**

10 Living cells could not function without enzyme-controlled reactions.

10.1 Explain how changing pH affects the rate of an enzyme-controlled reaction. **[3 marks]**

Exam Tip

Use example pHs: "at low pH...", and "at high pH..."

10.2 The enzyme trypsin breaks down casein (a form of protein) in milk. Give the name of the group of digestive enzymes that trypsin belongs to. **[1 mark]**

10.3 Trypsin breaks down casein, changing its colour from white to clear. Some scientists took a range of milk samples and mixed them with trypsin at different temperatures. They measured the rate at which trypsin breaks down casein using a spectrophotometer.

A spectrophotometer measures the amount of light transmitted through the liquid.

Suggest a method, using the spectrophotometer, to determine the optimum temperature for trypsin action. **[4 marks]**

10.4 The scientists noticed that the glass of the test tube containing the milk solution was cloudy.

Suggest and explain the effect of the clouded glass on the scientists' results. **[3 marks]**

11 Large multicellular organisms require systems to exchange gases efficiently.

11.1 Select the statement that best explains why single-celled organisms do not require gas exchange organs. **[1 mark]**

Single-celled organisms have a small surface area-to-volume ratio.

Single-celled organisms have a large surface area-to-volume ratio.

Single-celled organisms have a small surface area.

Single-celled organisms have a large surface area.

11.2 Explain the changes that occur in the body causing air to be drawn into the lungs. **[4 marks]**

11.3 Explain how the alveoli are adapted for gas exchange. **[3 marks]**

12 **Figure 8** shows an image of a blood smear viewed under a microscope.

Figure 8

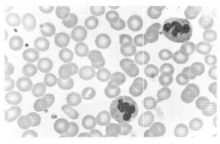

> ! **Exam Tip**
>
> Think about the function of a capillary.

12.1 Describe what conclusions you can draw about the composition of blood from **Figure 8**. **[3 marks]**

12.2 Pathologists often work with human blood samples to identify the presence of a disease. Evaluate the risks of working with blood products. Describe what precautions should be taken. **[3 marks]**

12.3 The average diameter of a human red blood cell is 0.008 mm. A pathologist measured the diameter of one of the red blood cells in **Figure 8** to be 10 mm in diameter. Calculate the magnification used to produce the blood smear image. **[2 marks]**

> ! **Exam Tip**
>
> The first step is to write down the equation you're going to use.

12.4 Some capillaries in the human body have an internal diameter of 0.01 mm. Calculate how many red blood cells are able to pass through a capillary at any one time and explain why this is advantageous. **[4 marks]**

13 The small intestine is covered in villi. A diagram of a villus is shown in **Figure 9**.

Figure 9

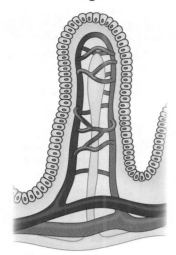

13.1 Identify which type of structure a villus is classified as. **[1 mark]**

cell   organ   tissue   organ system

13.2 Use information in **Figure 9** and your own knowledge to explain how a villus is adapted to its function. **[3 marks]**

13.3 Explain why villi cells have a large number of mitochondria. **[2 marks]**

13.4 The cells of the epithelial membrane are replaced every five to seven days. Describe the process by which this occurs. **[4 marks]**

14 Cyanide is used in a number of industrial processes and is also found in cigarette smoke. Cyanide interferes with respiration in cells. **Table 4** compares the rate of absorption of sugar in the small intestine of a healthy person, and the small intestine of a person who has been exposed to cyanide gas.

> **! Exam Tip**
>
> Cyanide is just a new context for content that you already know.

Table 4

	Healthy small intestine	Small intestine exposed to cyanide gas
relative rate of absorption	0.75	0.40

14.1 Calculate the percentage decrease in the rate of sugar absorption in the person who has inhaled cyanide gas. **[2 marks]**

14.2 Explain the differences in the results shown with reference to cell transport. **[6 marks]**

B6 Organisation in plants

Tissues in leaves

Leaves are organs because they contain many tissues that work together to perform photosynthesis.

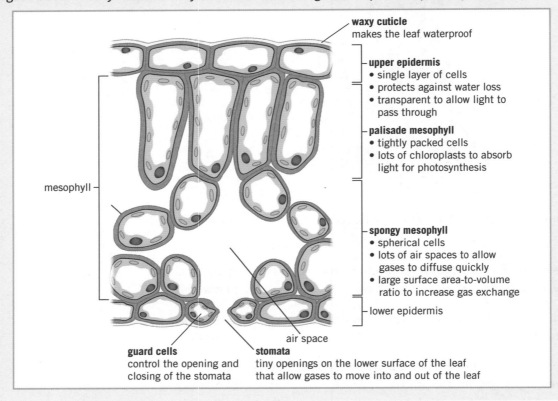

waxy cuticle
makes the leaf waterproof

upper epidermis
- single layer of cells
- protects against water loss
- transparent to allow light to pass through

palisade mesophyll
- tightly packed cells
- lots of chloroplasts to absorb light for photosynthesis

spongy mesophyll
- spherical cells
- lots of air spaces to allow gases to diffuse quickly
- large surface area-to-volume ratio to increase gas exchange

lower epidermis

mesophyll

air space

guard cells
control the opening and closing of the stomata

stomata
tiny openings on the lower surface of the leaf that allow gases to move into and out of the leaf

Stomata

Stomata are tiny openings in the undersides of leaves – this placement reduces water loss through evaporation.

They control gas exchange and water loss from leaves by:
- allowing diffusion of carbon dioxide into the plant for photosynthesis
- allowing diffusion of oxygen out of the plant.

Guard cells are used to open and close the stomata.

When a plant has plenty of water, the guard cells become turgid. The cell wall on the inner surface is very thick, so it cannot stretch as much as the outer surface. So as the guard cells swell up, they curve away from each other, opening the stoma.

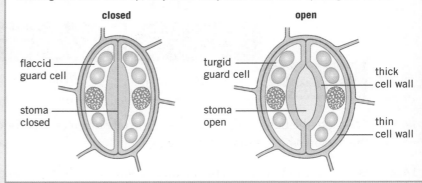

closed

flaccid guard cell

stoma closed

open

turgid guard cell

thick cell wall

stoma open

thin cell wall

Make sure you can write a definition for these key terms.

cuticle	epidermis	flaccid	guard cell	mesophyll	phloem	stomata
translocation	transpiration	transpiration stream		turgid	xylem	

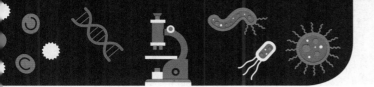

Transpiration

Description

Water is lost through the stomata by evaporation. This pulls water up from the roots through the **xylem** and is called transpiration. The constant movement of water up the plant is called the **transpiration stream**.

Importance

- provides water to cells to keep them **turgid**
- provides water to cells for photosynthesis
- transports mineral ions to leaves

Specialised tissues

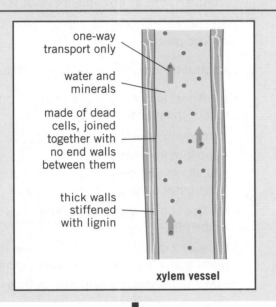

- one-way transport only
- water and minerals
- made of dead cells, joined together with no end walls between them
- thick walls stiffened with lignin

xylem vessel

Translocation

The movement of dissolved sugars from the leaves to the rest of the plant through the **phloem**.

- moves dissolved sugars made in the leaves during photosynthesis to other parts of the plant
- this allows for respiration, growth, and glucose storage

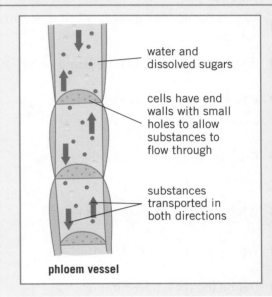

- water and dissolved sugars
- cells have end walls with small holes to allow substances to flow through
- substances transported in both directions

phloem vessel

Factors affecting the rate of transpiration

Factor	Effect on transpiration	Because...
temperature	higher temperatures *increase* the rate of transpiration	water evaporates faster in higher temperatures
humidity	lower humidity *increases* the rate of transpiration	the drier the air, the steeper the concentration gradient of water molecules between the air and leaf
wind speed	more wind *increases* the rate of transpiration	wind removes the water vapour quickly, maintaining a steeper concentration gradient
light intensity	higher light intensity *increases* the rate of transpiration	stomata open wider to let more carbon dioxide into the leaf for photosynthesis

Root hair cells

- increase absorption of water and mineral ions into the root by increasing the root surface area

- contain lots of mitochondria to transfer energy, which is used to take in mineral ions by active transport

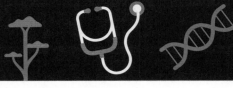
Learn the answers to the questions below then cover the answers column with a piece of paper and write as many as you can. Check and repeat.

B6 questions	Answers
1 Why is a leaf an organ?	there are many tissues inside the leaf that work together to perform photosynthesis
2 How is the upper epidermis adapted for its function?	• single layer of transparent cells allow light to pass through • cells secrete a waxy substance that makes leaves waterproof
3 How is the palisade mesophyll adapted for its function?	tightly packed cells with lots of chloroplasts to absorb as much light as possible for photosynthesis
4 How is the spongy mesophyll adapted for its function?	air spaces increase the surface area and allow gases to diffuse quickly
5 What is the function of the guard cells?	control the opening and closing of the stomata
6 What is the function of the xylem?	transport water and mineral ions from the roots to the rest of the plant
7 Give three adaptations of the xylem.	• made of dead cells • no end wall between cells • walls strengthened by a chemical called lignin to withstand the pressure of the water
8 What is the function of the phloem?	transport dissolved sugars from the leaves to the rest of the plant
9 What is the purpose of translocation?	transport dissolved sugars from the leaves to other parts of the plant for respiration, growth, and storage
10 Define the term transpiration.	movement of water from the roots to the leaves through the xylem
11 What is the purpose of transpiration?	• provide water to keep cells turgid • provide water to cells for photosynthesis • transport mineral ions to leaves
12 Name four factors that affect the rate of transpiration.	temperature, light intensity, humidity, and wind speed
13 What effect does temperature have on the rate of transpiration?	higher temperatures increase the rate of transpiration
14 What effect does humidity have on the rate of transpiration?	higher levels of humidity decrease the rate of transpiration
15 Why does increased light intensity increase the rate of transpiration?	stomata open wider to let more carbon dioxide into the leaf for photosynthesis
16 What is the function of the stomata?	allow diffusion of gases into and out of the plant
17 Where are most stomata found?	underside of leaves
18 What is the advantage to the plant of having a high number of stomata at this location?	reduces the amount of water loss through evaporation

Put paper here

Now go back and use the questions below to check your knowledge from previous chapters.

B6

Previous questions | Answers

Previous questions		Answers
List the structures air passes through when breathing in	Put paper here	mouth/nose → trachea → bronchi → bronchioles → alveoli
Give one advantage of using therapeutic cloning.	Put paper here	stem cells from the embryo are not rejected when transplanted because they have the same genes as the patient
How does the structure of a vein relate to its function?	Put paper here	carries blood back to the heart at low pressure – doesn't need thick, elasticated walls, but has valves to prevent blood flowing the wrong way
What does denatured mean?	Put paper here	shape of an enzyme's active site is changed by high temperatures or an extreme pH, so it can no longer bind with the substrate
How are villi adapted for exchanging substances?	Put paper here	• long and thin – increases surface area • one-cell-thick membrane – short diffusion pathway • good blood supply – maintains a steep concentration gradient

Maths Skills

Practise your maths skills using the worked example and practice questions below.

Calculating rate of transpiration	Worked Example	Practice

Calculating rate of transpiration

Transpiration cannot be measured directly. Instead it is determined by measuring the decrease in mass of a plant due to water loss, or by measuring the volume of water absorbed by the plant.

A **potometer** can be used to determine the rate of transpiration by measuring the volume of water absorbed by a plant.

The volume of water absorbed can be calculated by measuring the distance travelled by an air bubble in a given time in the potometer. The faster the bubble moves, the greater the rate of water uptake, and the greater the assumed rate of transpiration.

Worked Example

A group of students used a potometer to measure the volume of water absorbed by a plant under three different conditions over 25 minutes. Their results were:

- normal conditions: 2.4 ml water absorbed
- high temperature: 3.1 ml water absorbed
- low humidity: 3.5 ml water absorbed

Work out the transpiration rate of the plant under each condition.

transpiration rate (ml/min)

$$= \frac{\text{volume of water absorbed (ml)}}{\text{time (min)}}$$

- normal conditions: $\frac{2.4}{25}$
  $= 0.096$ ml/min
- high temperatures:
  $\frac{3.1}{25} = 0.124$ ml/min
- low humidity: $\frac{3.5}{25} = 0.140$ ml/min

Practice

1 The table below shows the volume of water absorbed by a plant under three different conditions in 10 minutes. Calculate the transpiration rate for the plant under each condition.

Conditions	Volume of water in ml	Time in mins	Transpiration rate in ml/min
normal	1.1	10	
high temperature	1.3	10	
low humidity	1.5	10	

2 Which condition produced the highest transpiration rate? Explain this result.

3 How would you expect the volume of water absorbed to differ to that under normal conditions if a fan was set up to blow air over the plant?

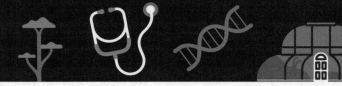

Practice

Exam-style questions

01 Four leaves of approximately the same size were removed from an oak tree.

Petroleum jelly was spread over the surface of leaves **A–C**. This acts as a waterproof agent to prevent water loss from the leaves.

All four leaves were then hung from a piece of string.

Figure 1

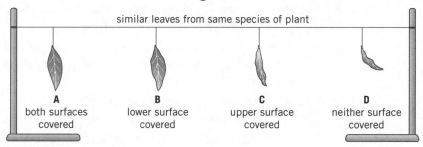

01.1 Suggest why no petroleum jelly was spread over the surface of leaf **D**. **[1 mark]**

it a controll release . test.

01.2 The mass of each leaf was weighed at regular intervals.

Name the apparatus used to measure the mass of each leaf. **[1 mark]**

top pan Balloon .

01.3 The results of the investigation are shown in **Figure 2**.

Figure 2

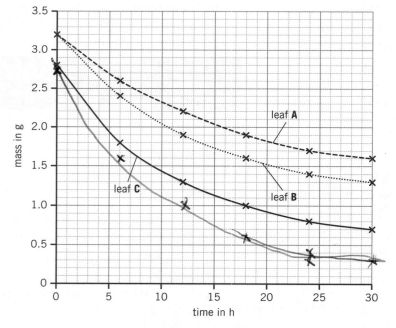

The values for leaf **D** are shown in **Table 1**.

Exam Tip

You can use the points already plotted as a guide to find 6 and 12 hours.

Table 1

Leaf	Mass in g					
	start	after 6 h	after 12 h	after 18 h	after 24 h	after 30 h
D	2.7	1.6	1.0	0.6	0.4	0.3

Plot these data on **Figure 2**. Draw a line of best fit. **[3 marks]**

01.4   Identify which leaf lost water the fastest. Give a reason for your answer. **[2 marks]**

leaf plant value fastest because it ⟶

_____

02   **Figure 3** shows a cross-section through a leaf.

Figure 3

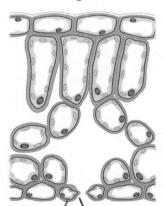

02.1   Identify and label a stoma on the diagram. **[1 mark]**

02.2   Describe the role of guard cells in controlling water loss from a leaf. **[2 marks]**

Because turgid to dull stomata when water is low and flaccid to close vapor when water's low self regulating ⟶

02.3   A scientist calculated that the width of the cross-section they were viewing through the microscope was 250 µm.

Estimate the width of the guard cell. **[1 mark]**

 _____ µm

> ⚠ **Exam Tip**
>
> Start by measuring the width of the cross-section, then measure the width of the guard cell.

03 A group of students were studying the factors that affect the rate of transpiration in a plant. They used a potometer (**Figure 4**) to measure the rate of water uptake by a plant. The rate of water uptake can be used as an approximation of the rate of transpiration in a plant.

Figure 4

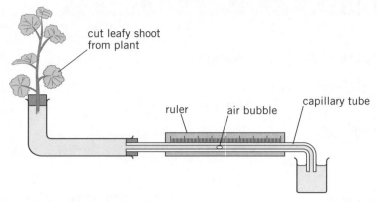

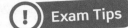

! Exam Tips

You may not have seen a practical set up like this before, but don't panic!

This question is testing if you can apply your practical skills.

03.1 Define the term transpiration. **[2 marks]**

03.2 The students studied the rate of water uptake of the plant by measuring the distance travelled by an air bubble at 5-minute intervals over 30 minutes. Explain why the rate of water uptake and the rate of transpiration are different. **[2 marks]**

03.3 The students' results are shown in **Table 2**.

Table 2

Time in min	Distance moved by bubble from start point in mm
0	0
5	4
10	8
15	11
20	16
25	19
30	25

Plot a graph of the water uptake of the plant on the axes below. Draw a suitable line of best fit. **[3 marks]**

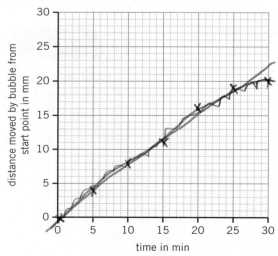

03.4 Describe the relationship between the distance moved by the bubble and time. **[1 mark]**

Exam Tip

Use data from the graph.

03.5 Use your graph to determine how long it took the bubble to move 10 mm. **[1 mark]**

03.6 Calculate the rate of water uptake by the plant between 10 and 20 min. **[4 marks]**

Exam Tip

Show your working on the graph.

03.7 The experiment was carried out at room temperature. Explain how you would expect the graph to differ if the experiment was repeated at 35°C. **[2 marks]**

Exam Tip

This is a maths skill you might use more often in chemistry, but you should be prepared for it to come up anywhere in science.

04 Plants use transport systems to move materials around inside them.

04.1 Name the structure found in stems that contains the xylem and phloem tissue. **[1 mark]**

04.2 Define the term translocation. **[1 mark]**

04.3 Describe the main structural differences between xylem and phloem tissue. **[4 marks]**

Exam Tip

You might not have covered greenfly in class, but apply what you know from your lessons to this new context.

04.4 Suggest why plant pests such as greenfly bite into the phloem tissue in a plant. **[1 mark]**

05 A student carried out an investigation on beech leaves to compare the number of stomata present on the upper and lower surfaces of the leaf.

05.1 Describe how the student could take samples from the leaf to count the number of stomata present. **[3 marks]**

05.2 The student's results are shown in **Table 3**.

Table 3

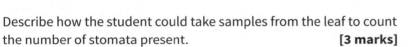

Surface	Number of stomata present					Mean
	Sample 1	Sample 2	Sample 3	Sample 4	Sample 5	
Upper	1	2	2	3	2	2
Lower	36	42	35	41	37	

Calculate the mean result for the lower surface of the leaf. Give your answer to **two** significant figures. **[2 marks]**

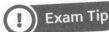
Exam Tip

Remember significant figures and decimal places are different things. Don't get them confused!

05.3 The student concluded that most stomata are found on the lower surface of a beech leaf. Explain why this is an advantage for a beech tree. **[2 marks]**

06 **Figure 5** shows a plant cell.

Figure 5

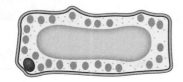

06.1 Identify which part of a plant the cell has been taken from. Choose **one** answer. **[1 mark]**

palisade mesophyll   phloem tissue   root   xylem tissue

06.2 Give a reason for your answer to **06.1**. **[1 mark]**

06.3 Describe how water moves between cells in a leaf. **[2 marks]**

06.4 The main function of a leaf is to perform photosynthesis to provide food for the plant. Describe how the tissues inside a leaf are organised to maximise photosynthesis. **[6 marks]**

07 As well as anchoring a plant into the ground, roots are also responsible for the uptake of water and mineral ions from the soil.

07.1 Explain how a root hair cell is adapted for the uptake of water and mineral ions. **[3 marks]**

07.2 **Figure 6** represents the movement of water and mineral ions into the root hair cell.

Figure 6

outside cell   inside cell

cell membrane

high concentration of substance A   → **process X** →   low concentration of substance A

low concentration of substance B   → **process Y** →   high concentration of substance B

Identify and name the process in **Figure 6** that represents the uptake of mineral ions. Give a reason for your answer. **[2 marks]**

07.3 Name the vessel that transports mineral ions around the plant. **[1 mark]**

07.4 Describe **one** use of mineral ions in a plant. **[1 mark]**

08 Scientists can use sampling and counting techniques to investigate the distribution of stomata on leaves. **Figure 7** is an observational diagram produced by a scientist when looking at a lower leaf epidermis. For each sample observed, the scientist calculated the density of stomata present in the form: *number of stomata per mm²*. Partially visible stomata were counted as present.

Figure 7

0.40 mm

08.1 Calculate the density of stomata for the sample shown in **Figure 7**. **[5 marks]**

Exam Tips

This is a six mark question, so make sure you write enough.

Think about how the cells in a plant differ depending on their location and function.

Exam Tips

'Explain' questions are asking *why*.

An answer describing what a root hair cells looks like won't get marks.

Exam Tips

Try to get into the mind of the examiner.

This is a 2 mark question – the first mark will be for identifying the process and the second mark will be for the reason.

Exam Tip

You'll need to work out the area of the circle first.

08.2 The scientist then estimated the total surface area of the leaf from which the sample was taken. Suggest how the scientist estimated the leaf surface area. **[2 marks]**

08.3 The scientist measured the leaf's surface area to be approximately 8 cm². Estimate the number of stomata that would be found on the surface of this leaf. **[3 marks]**

08.4 Suggest and explain how the scientist's results may have been different if the sample was taken from the upper surface of the leaf. **[3 marks]**

09 Marram grass grows on sand dunes. Sand dunes are generally very dry and windy habitats. The leaves of marram grass are specially adapted to reduce water loss by transpiration. Some of these features are shown in **Figure 8**.

Figure 8

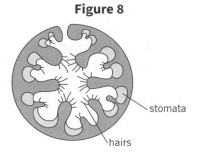

stomata

hairs

09.1 Define the term transpiration. **[2 marks]**

09.2 Identify which property of the waxy cuticle reduces the rate of transpiration. Choose **one** answer. **[1 mark]**

impermeable   large surface area   reflective   thermal insulator

09.3 Using **Figure 8**, suggest and explain how the rolled leaves, stomata, and leaf hairs work together to reduce the rate of transpiration. **[3 marks]**

> ! **Exam Tip**
>
> You'll need to refer to rolled leaves, stomata, and leaf hairs if you want to get full marks on this question.

10 The photograph in **Figure 9** was taken through a microscope. It shows a vascular bundle in a leaf. Vascular bundles contain both xylem and phloem tissue.

Figure 9

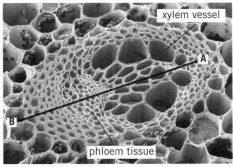

xylem vessel

A

B

phloem tissue

10.1 Line **A–B** shows the width of the vascular bundle. Vascular bundles in this species of plant have a mean width of 250 μm. Calculate the magnification of the image. **[3 marks]**

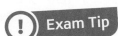

> ! **Exam Tip**
>
> Step one is to write down the equation for magnification.

10.2 Describe one difference between the structure of the xylem vessel and the phloem tube in **Figure 9**. **[1 mark]**

10.3 Name the chemical present in xylem vessel walls that provides the strength to withstand the pressure of the movement of water in the plant. **[1 mark]**

10.4 The chemical named in **10.3** can be seen using a stain. Use this information to plan how you could find the position of the vascular bundles in a stem. **[3 marks]**

10.5 Deer are a concern in managed woodlands. They eat tree bark and new tree shoots, and rub their antlers on tree trunks to leave a scent marker to warn other deer away from the area. Explain the reasons why protective collars are placed around tree saplings in areas of managed woodland. **[6 marks]**

11 Dementia is a condition that causes a decline in brain function. It affects around 850 000 people in the United Kingdom. Symptoms include memory loss, difficulties with movement, and speech problems. Stem cells are being investigated as a possible cure for dementia. Discuss the arguments for and against using stem cell research to find a cure for dementia. **[6 marks]**

12 **Figure 10** shows a bacterial cell.

Figure 10

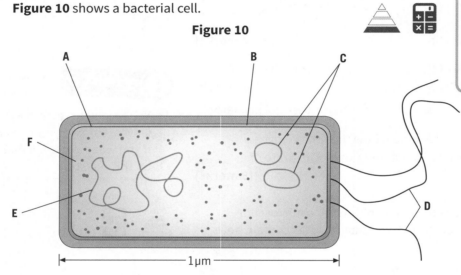

12.1 Identify which part of the cell (**A**–**F**) is the cell membrane. **[1 mark]**

12.2 Describe the function of structure **D**. **[1 mark]**

12.3 Use **Figure 10** to explain how you can tell that this is a prokaryotic cell. **[2 marks]**

12.4 Calculate the magnification of the image in **Figure 10**. **[2 marks]**

13 A group of students were asked to investigate the effect of sugar solutions of different concentrations on samples of strawberry tissue.

Figure 11

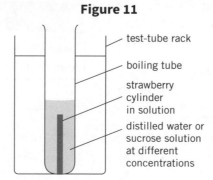

test-tube rack

boiling tube

strawberry cylinder in solution

distilled water or sucrose solution at different concentrations

13.1 Using the apparatus in **Figure 11** and other standard laboratory equipment, suggest a method the students could have used to accurately determine the concentration of sucrose in the strawberry samples. **[6 marks]**

13.2 To increase the size of strawberries from their plants, growers provide the plants with lots of water. Discuss the potential economic advantages and disadvantages to the grower of providing excess water to strawberry plants. **[6 marks]**

> **! Exam Tip**
>
> You'll need to cover *both* the advantages and disadvantages to get full marks here.

14 Fish and single-celled amoebas are both examples of living organisms found in water.

14.1 Explain which of these organisms has a specialised transport system. **[3 marks]**

14.2 Humans have a double circulatory system to transport materials around the body. Frogs have only a partial double circulatory system as shown in **Figure 12**.

Figure 12

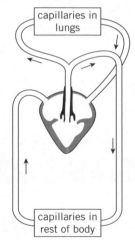

> **! Exam Tip**
>
> Use this space to sketch out the human circulatory system, so it's easier for you to compare them.

Using **Figure 12** and your own knowledge, give **two** similarities and **two** differences between the frog and human circulatory systems. **[4 marks]**

14.3 Suggest and explain why the frog transport system is less effective in supplying body tissues with oxygen. **[3 marks]**

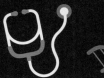
B7 The spread of diseases

Viruses live and reproduce rapidly inside an organism's cells. This can damage or destroy the cells.

Viruses

	Spread by	Symptoms
measles	inhalation of droplets produced by infected people when sneezing and coughing	• fever • red skin rash • complications can be fatal – young children are vaccinated to immunise them against measles
HIV (human immunodeficiency virus)	• sexual contact • exchange of body fluids (e.g., blood when drug users share needles)	• flu-like symptoms at first • virus attacks the body's immune cells, which can lead to AIDS – where the immune system is so damaged that it cannot fight off infections or cancers
TMV (tobacco mosaic virus – plants)	• direct contact of plants with infected plant material • animal and plant vectors • soil: the pathogen can remain in soil for decades	• mosaic pattern of discolouration on the leaves – where chlorophyll is destroyed • reduces plant's ability to photosynthesise, affecting growth

Pathogens

Microorganisms that cause disease are called **pathogens**.

There are four types of pathogen:
• bacteria
• fungi
• protists
• viruses.

Pathogens can be spread:
• in the air
• in water
• through direct contact.

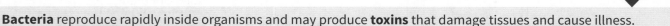

Bacteria reproduce rapidly inside organisms and may produce **toxins** that damage tissues and cause illness.

Bacteria

	Spread by	Symptoms	Prevention and treatment
Salmonella	bacteria in or on food that is being ingested	Salmonella bacteria and the toxins they produce cause • fever • abdominal cramps • vomiting • diarrhoea	poultry are vaccinated against Salmonella bacteria to control spread
gonorrhoea	direct sexual contact – gonorrhoea is a **sexually transmitted disease** (STD)	• thick yellow or green discharge from the vagina or penis • pain when urinating	• treatment with antibiotics (many antibiotic-resistant strains have appeared) • barrier methods of contraception, such as condoms

Controlling the spread of communicable disease

There are a number of ways to help prevent the spread of communicable diseases from one organism to another.

Hygiene
Hand washing, disinfecting surfaces and machinery, keeping raw meat separate, covering mouth when coughing/sneezing, etc.

Isolation
Isolation of infected individuals – people, animals, and plants can be isolated to stop the spread of disease.

Controlling vectors
If a vector spreads a disease destroying or controlling the population of the vector can limit the spread of disease.

Vaccination
Vaccination can protect large numbers of individuals against diseases.

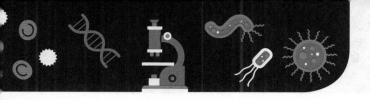

B7

Fungi

	Spread by	Symptoms	Prevention and treatment
rose black spot	water and wind	• purple or black spots on leaves, which turn yellow and drop early • reduces plant's ability to photosynthesise, affecting growth	• **fungicides** • affected leaves removed and destroyed

Protists

	Spread by	Symptoms	Prevention and treatment
malaria	mosquitos feed on the blood of infected people and spread the protist pathogen when they feed on another person – organisms that spread disease by carrying pathogens between people are called **vectors**	• recurrent episodes of fever • can be fatal	• prevent mosquito vectors breeding • mosquito nets to prevent bites • anti-malarial medicine

Detection and identification of plant diseases

Signs that a plant is diseased
- stunted growth
- spots on leaves
- areas of rot or decay
- growths
- malformed stems or leaves
- discolouration
- pest infestation

Ways of identifying plant diseases
- gardening manuals and websites
- laboratory testing of infected plants
- testing kits containing monoclonal antibodies (Chapter 9 *Monoclonal antibodies*)

Plant defences

Physical barriers
- cellulose cell walls – provide a barrier to infection
- tough waxy cuticle on leaves
- bark on trees – a layer of dead cells that can fall off

Chemical barriers
- many plants produce antibacterial chemicals
- poison production stops animals eating plants

Mechanical adaptations
- thorns and hairs stop animals eating plants
- leaves that droop or curl when touched to scare herbivores or dislodge insects
- some plants **mimic** the appearance of unhealthy or poisonous plants to deter insects or herbivores

Plant diseases and insects

Plant diseases can also be directly caused by insects.

Aphids are insects that suck sap from the stems of plants. This results in
- reduced rate of growth
- wilting
- discolouration of leaves.

Ladybirds can be used to control aphid infestations as ladybird larvae eat aphids.

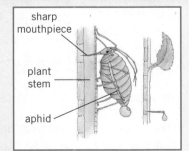
sharp mouthpiece
plant stem
aphid

Key terms

Make sure you can write a definition for these key terms.

aphid	bacterium	communicable disease	fungicide	fungus	mimic
pathogen	protist	sexually transmitted disease (STD)	toxin	vector	virus

B7 Knowledge 73

 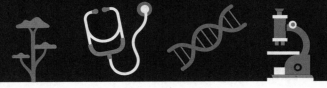

Learn the answers to the questions below then cover the answers column with
a piece of paper and write as many as you can. Check and repeat.

B7 questions		Answers
1	What is a communicable disease?	a disease that can be transmitted from one organism to another
2	What is a pathogen?	a <u>microorganism</u> that causes disease
3	Name four types of pathogen.	bacteria, <u>fungi</u>, protists, viruses
4	How can pathogens spread?	air, water, direct contact
5	How do bacteria make you ill?	produce toxins that damage <u>tissues</u>
6	How do viruses make you ill?	reproduce rapidly inside cells, damaging or destroying them
7	Name three examples of viral diseases.	<u>measles</u>, HIV, tobacco mosaic virus
8	Name two examples of bacterial diseases.	*Salmonella*, gonorrhoea
9	Name four methods of controlling the spread of communicable disease.	good hygiene, <u>isolating infected individuals</u>, controlling vectors, vaccination
10	Describe an example of a protist disease.	malaria – caused by <u>a protist pathogen</u> that is spread from person to person by mosquito bites, and causes recurrent fevers
11	Describe an example of a fungal disease in plants.	rose black spot – spread by <u>water and wind</u>, and affects plant growth by reducing a plant's ability to photosynthesise
12	How can the cause of a plant disease be identified?	<u>gardening manuals and websites</u>, laboratory testing, <u>monoclonal antibody kits</u>
13	What are three mechanical defences that protect plants?	thorns and hairs, <u>leaves that droop or curl</u>, mimicry to trick animals
14	Give three physical defences of plants.	<u>cellulose cell walls</u>, tough waxy cuticles, bark on trees
15	How can aphids be controlled by gardeners?	introduce ladybirds to eat the aphids
16	How can plant diseases be detected?	areas of decay, discolouration, growths, malformed stems or leaves, presence of pests, spots on leaves, and stunted growth

Put paper here

(!) Exam Tip

Question 7 and question 10 have different command words.
That's why the answers are different lengths. For a 'name'
question, just a few words are needed, whereas for a
'describe' question you need to say what something is like.

Now go back and use the questions below to check your knowledge from previous chapters.

B7

Previous questions

Answers

Previous questions	Answers
Why are enzymes described as specific?	each enzyme only catalyses a specific reaction, because the active site only fits together with certain substrates (like a lock and key)
How is the palisade mesophyll adapted for its function?	tightly packed cells with lots of chloroplasts to absorb as much light as possible for photosynthesis
What is the function of a root hair cell?	absorbs minerals and water from the soil
What is cell division by mitosis?	body cells divide to form two identical daughter cells
What is the function of the phloem?	transport dissolved sugars from the leaves to the rest of the plant
Where are adult stem cells found?	bone marrow
What is the purpose of transpiration?	• provide water to keep cells turgid • provide water to cells for photosynthesis • transport mineral ions to leaves
Describe the effect of pH on enzyme activity.	different enzymes have a different optimum pH at which their activity is greatest – at a pH much lower or higher than this enzyme activity decreases and stops

Put paper here

Maths Skills

Practise your maths skills using the worked example and practice questions below.

Calculating percentage change	Worked Example	Practice
To calculate percentage change you need to work out the difference between the two numbers you are comparing. Then, you divide the difference by the original number and multiply the answer by 100. If your answer is a negative number, this equals a percentage decrease. percentage change = $\dfrac{\text{difference}}{\text{original number}} \times 100$	In 2009, the number of deaths in England caused by MRSA was 800. In 2010, the number of deaths had fallen to 500. Calculate the percentage change in the number of deaths caused by MRSA between 2009 and 2010. Work out the difference in the two numbers you are comparing: $800 - 500 = 300$ Divide the difference (300) by the original number: $\dfrac{300}{800} = 0.375$ Multiply by 100: $0.375 \times 100 = 37.5$ Percentage change in deaths caused by MRSA = 37.5%	The table below gives information about the number of deaths per year in England from MRSA and *Clostridium difficile* over four years. | Year	MRSA	C. difficile
2007	1800	8100
2008	1730	5300
2009	800	3890
2010	500	4570 | 1 Calculate the percentage change in deaths caused by MRSA from 2007 to 2008. 2 Calculate the percentage change in deaths caused by *C. difficile* from 2007 to 2008. 3 Calculate the percentage change in deaths caused by *C. difficile* from 2009 to 2010.

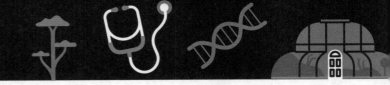

Practice

Exam-style questions

01 **Figure 1** shows a simplified image of human immunodeficiency virus (HIV).

Figure 1

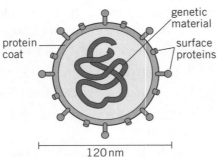

protein coat

genetic material

surface proteins

120 nm

01.1 Give **two** ways this virus can be spread between people. **[2 marks]**

1 _sexual contact._

2 _bodily fluids._

01.2 Using the information from **Figure 1**, explain how you can tell this is a viral cell and not a bacterial cell. **[4 marks]**

smaller than a bacterial cell

protein coat by

01.3 Describe how a virus spreads within the body. **[4 marks]**

a virus enter a body cell were is replicing
and doing the cell energy.

01.4 HIV infects white blood cells, which measure approximately 15 μm in diameter.

Calculate how many times larger a white blood cell is than a HIV particle.

Give your answer in terms of order of magnitude. **[4 marks]**

_15 × 100 = 1500 nm_

_1500 ÷ 120 =_

_12.5_

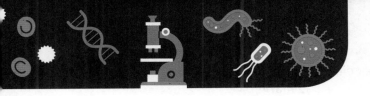
01.5 In 2005, there were 8000 new cases of HIV in the UK. By 2016 this figure had fallen to 4300. Suggest and explain **two** reasons why the number of new cases of HIV has fallen by such a significant amount. **[4 marks]**

Better public awareness meaning that more people prepare safe sex.
crackdown on drugs reducing the transfere rate from sharing needles

02 **Figure 2** shows a graph of the number of bacteria present in a colony over time, under optimum conditions for reproduction.

Figure 2

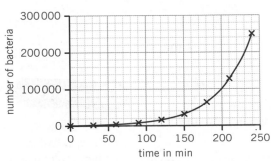

02.1 Describe the trend shown in the graph. **[2 marks]**

exponential prative.

02.2 Name the process by which bacteria reproduce. **[1 mark]**

~~sexual~~ ~~repro~~ asexual reproduction

02.3 Identify the time at which there were 100 000 bacteria present in the sample. **[1 mark]**

200 min.

02.4 Using **Figure 2** and your own knowledge of bacterial reproduction, calculate the time from the start of the experiment at which 1 000 000 bacteria would be present in the sample. **[4 marks]**

2510×4^2

$40 000$

02.5 Suggest and explain how the graph would appear different if the bacteria were kept at a much lower temperature. **[2 marks]**

the graph would be a lower frequency

03 Diseases are caused by different types of pathogen.

03.1 Draw **one** line from each disease to the correct pathogen type that causes the disease. **[3 marks]**

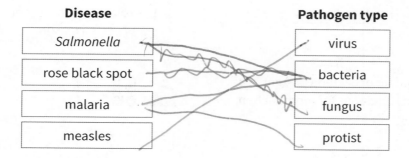

Disease — Pathogen type

Salmonella — virus
rose black spot — bacteria
malaria — fungus
measles — protist

Exam Tip

Only draw one line to and from each box. Any more will mean you don't get the marks, even if one of the lines is correct.

03.2 Give **one** symptom caused by measles. **[1 mark]**

red rash

03.3 Describe how measles is spread. **[1 mark]**

in spit droplets on breath / coughs

03.4 In 2018, there were 14.4 reported cases of measles per million people in the United Kingdom. The UK population in 2018 was 66 million.

Calculate the number of cases of measles in the UK in 2018. **[1 mark]**

950

Exam Tip

This is only a 1 mark question – if the way to answer it isn't obvious then you're over-thinking it!

03.5 In Ukraine in 2018, the number of cases of measles was 1204 per million people.

Calculate how many more cases of measles there were per million people in the Ukraine compared to the UK in 2018. **[1 mark]**

1189.6

Exam Tip

Use data and examples in your answer.

04 Malaria is a disease that kills over 400 000 people worldwide each year.

04.1 Give the type of pathogen that causes malaria. **[1 mark]**

Protist

04.2 Name the vector that is responsible for transmitting malaria. **[1 mark]**

Mosquitos

04.3 Sub-Saharan Africa is known as a malarial hotspot.

Discuss the steps a traveller should take when visiting this region to minimise the risk of contracting a fatal malarial infection. **[6 marks]**

sleep in a bug net

05 Diseases can often be recognised by the symptoms they cause.

05.1 Draw **one** line from each disease to the correct symptom. **[3 marks]**

Disease		Symptoms
tobacco mosaic virus		purple or black spots on leaves
gonorrhoea		discolouration of leaves
rose black spot		fever, vomiting, diarrhoea
Salmonella		yellow or green discharge from sexual organs

05.2 Complete the following sentences about sexually transmitted diseases. **[4 marks]**

_____gonorrhoea_____ and _____HIV_____ are two examples of sexually transmitted diseases. They are spread through sexual intercourse. Some forms of contraception, such as _____condoms_____ , are effective at preventing the spread of these types of diseases. This is because they provide a _____barrier_____ , preventing the pathogens being passed from one person to another.

05.3 A school suffers from an outbreak of whooping cough, an infectious disease spread by droplet infections or direct contact with an infected person or contaminated surface.

Tick **three** steps that would help the school to control the spread of the infection. **[3 marks]**

✓	Send infected children home.
✓	Prevent visitors from coming into the school.
	Employ a new school nurse.
✓	Wash surfaces down with disinfectant.
	Teach students about the benefits of vaccination.

06 The spread of many diseases, such as the common cold and some forms of food poisoning, can be avoided through good hygiene practices.

06.1 Describe how each of the following approaches prevents the spread of pathogens: **[4 marks]**
- washing hands before preparing food *prevents bacteria entering food.*
- covering face when coughing or sneezing *stops moisture carrying pathogens*
- wiping down surfaces with disinfectant *pathogens are being spread.*
- isolation of infected people. *not picked up when people touch surface*

can't be transmitted

06.2 Some raw meat products contain pathogens. Suggest and explain **two** ways in which a restaurant could minimise the risk of infection for its customers. **[4 marks]**

06.3 *Salmonella* bacteria are sometimes found in uncooked chicken. Explain how the *Salmonella* bacteria cause you to feel unwell if you eat an infected meat product. **[3 marks]**

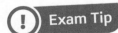
Exam Tip

Salmonella causes illness in the same way that other bacteria do.

06.4 Suggest and explain **one** way to stop the supply of infected meat to the UK market. **[2 marks]**

07 Rose black spot is the most serious disease to affect rose plants.

07.1 Name the type of pathogen that causes rose black spot disease. **[1 mark]**

07.2 Give **two** symptoms of rose black spot disease. **[2 marks]**

07.3 Explain why a plant infected with rose black spot disease does not grow properly. **[2 marks]**

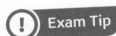
Exam Tip

You'll need to include details about what effects the disease has.

07.4 A commercial flower grower notices that several of his plants are infected with rose black spot disease.

Give **two** possible treatment options available to the grower to prevent further spread of the infection. **[2 marks]**

08 Around half of the human population live in regions that are exposed to malaria. In 2015, there were over 200 million cases of malaria worldwide.

08.1 Give **one** symptom of malaria. **[1 mark]**

08.2 Many people believe 'malaria is caused by mosquitos'. Explain why this statement is **not** correct. **[2 marks]**

08.3 **Figure 3** shows a graph of the number of deaths worldwide due to malaria between the years 2000 and 2015.

Exam Tip

Drawing a line of best fit can help you describe the trend.

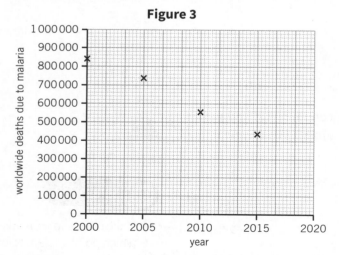

Figure 3

Describe the trend shown by the data. Suggest **two** reasons for this trend in the data. **[3 marks]**

08.4 Predict the number of worldwide deaths due to malaria in 2020. **[2 marks]**

Exam Tip

You'll need to draw on the graph to answer this.

08.5 Gross domestic product (GDP) is a measure of the wealth of a country. In general, the larger the GDP, the higher the standard of living in a country.

Figure 4 shows GDP data against the number of cases of malaria per 100 000 population for different countries.

Figure 4

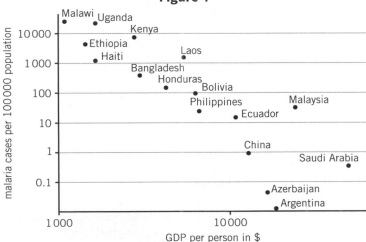

Describe the relationship shown by this scatter diagram. **[2 marks]**

Exam Tip

For the next two questions, you'll need to use examples from the graph.

08.6 Suggest and explain **two** reasons for this relationship. **[4 marks]**

09 The tobacco mosaic virus (TMV) infects tomato plants and causes a distinct mosaic pattern on the leaves.

09.1 Explain why this is an example of a communicable disease. **[2 marks]**

09.2 Explain the effects of the virus on the growth of a tomato plant. **[4 marks]**

Exam Tip

Think about the effects of the virus, and how these might have knock on effects.

09.3 Using your knowledge of TMV, identify which of the cell components TMV infects. Choose **one** answer. **[1 mark]**

nucleus   mitochondria   chloroplast   cell wall

09.4 TMV does not usually kill the plant it has infected, instead it significantly reduces the plant's yield. TMV is most commonly spread between plants by mechanical transmission – through workers' hands and tools. Some species of insects also act as vectors for TMV. TMV is also able to survive in soil for around 50 years.

Using the information above, suggest and explain **two** ways farmers could try to prevent the spread of TMV. **[4 marks]**

10   **Figure 5** shows a graph of the number of deaths worldwide due to HIV between 1990 and 2015.

Figure 5

10.1   Explain how an infection with HIV can lead to death.   **[3 marks]**

10.2   Name the type of drug used to treat HIV infections.   **[1 mark]**

10.3   Describe the worldwide trend in deaths linked to HIV between 1990 and 2015.   **[2 marks]**

10.4   Suggest and explain **one** reason for the trend between 2005 and 2015.   **[2 marks]**

10.5   Using information in **Figure 5**, calculate the number of worldwide deaths per 100 000 people due to HIV infection in 2008.   **[1 mark]**

10.6   Calculate the percentage increase in deaths linked to HIV between 1998 and 2005.   **[3 marks]**

(!) **Exam Tip**

This graph changes direction halfway. Make sure you refer to this in your answer, and use data to back up what you say.

11   Rose black spot is a disease that affects roses. It leads to yellow and black patterns on the leaves.

11.1   Identify the type of pathogen that causes rose black spot.   **[1 mark]**

   fungus   virus   bacteria   protist

11.2   Identify the correct chemical that can be used to treat rose black spot.   **[1 mark]**

   insecticide   herbicide   fungicide   pesticide

12   To prevent insects damaging crops, some farmers spray their crops with chemicals.

12.1   Name the type of chemical that can be used to prevent aphids (greenfly) from attacking the crop.   **[1 mark]**

12.2   As well as eating crops, aphids can cause plant disease. Suggest and explain how this can occur.   **[2 marks]**

(!) **Exam Tip**

This question is asking about what the plant can do, not what things can be added to the plant to help protect it.

12.3 Suggest and explain **two** ways a plant could protect itself from
aphid attack. **[4 marks]**

12.4 A group of scientists investigated the effect of pesticide use on grain
yield. Their results are in **Table 1**.

Table 1

Pesticide application level, in % compared to national average	Grain yield in tonnes per hectare
0	5.0
50	7.0
100	7.5

Calculate the percentage change in the grain yield when no
pesticides are used, compared to the national average. **[2 marks]**

12.5 Using **Table 1**, write a conclusion for the effect of pesticides on crop
yields, giving reasons for your answer. **[4 marks]**

13 Duckweed is commonly found on the surface of ponds.

Table 2 shows the concentrations of some minerals in the
duckweed cells and in the surrounding pond water.

Table 2

	Ion concentration in M×10⁻³		
	Calcium	Sulfate	Potassium
plant cells	12.0	12.0	45.0
pond water	0.9	0.3	0.5

13.1 Identify and explain the process by which the duckweed would take
the minerals into its cells. **[3 marks]**

13.2 Duckweed has a number of features that make it highly adapted for
living in a pond. One of these features is air pockets inside the leaf
called aerenchyma.

Suggest how these, in combination with the roots, help the
plant to float. **[2 marks]**

13.3 Unlike most plants, the stomata in duckweed are found on the
upper surface of their leaves.

Suggest why this is an important adaptation. **[2 marks]**

13.4 Explain why controlling transpiration is not very important to
duckweed. **[2 marks]**

B8 Preventing and treating disease

Non-specific defences

Non-specific defences of the human body against all pathogens include:

Skin
- physical barrier to infection
- produces antimicrobial secretions
- microorganisms that normally live on the skin prevent pathogens growing

Nose

Cilia and **mucus** trap particles in the air, preventing them from entering the lungs.

Trachea and bronchi produce mucus, which is moved away from the lungs to the back of the throat by cilia, where it is expelled.

Stomach

Produces strong acid (pH 2) that destroys pathogens in mucus, food, and drinks.

White blood cells

If a pathogen enters the body, the immune system tries to destroy the pathogen.

The function of **white blood cells** is to fight pathogens.

There are two main types of white blood cell – lymphocytes and phagocytes.

> **Revision tip**
>
> Vaccines have saved countless lives since their development over two hundred years ago. Despite occasional concerns to the contrary, they are safe to use and highly effective.

Lymphocytes

Lymphocytes fight pathogens in two ways:

Antitoxins

Lymphocytes produce **antitoxins** that bind to the toxins produced by some pathogens (usually bacteria). This *neutralises* the toxins.

Antibodies

Lymphocytes produce **antibodies** that target and help to destroy specific pathogens by binding to **antigens** (proteins) on the pathogens' surfaces.

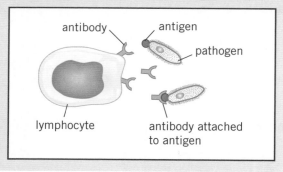

Phagocytes

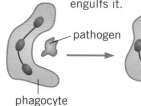

1 Phagocytes are attracted to an area of infection.
2 The phagocyte surrounds the pathogen and engulfs it.
3 Enzymes that digest and destroy the pathogen are released.

Herd immunity

If a large proportion of a population is vaccinated against a disease, the disease is less likely to spread, even if there are so unvaccinated individuals.

> **Revision tip**
>
> It's a common misconception that antibiotic resistance arises when people become resistant to a drug. In reality, it is the bacteria that evolve resistance.

🔑 Key terms

Make sure you can write a definition for these key terms.

antibiotic   antibody   antigen   antitoxin   cilia   dose   double-blind trial

efficacy   mucus   peer review   placebo   toxicity   vaccine   white blood cell

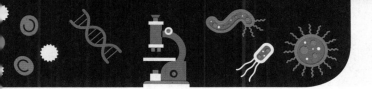

Treating diseases

Antibiotics

- Antibiotics are medicines that can kill *bacteria* in the body.
- Specific bacteria need to be treated by specific antibiotics.
- Antibiotics have greatly reduced deaths from infectious bacterial diseases, but antibiotic-resistant strains of bacteria are emerging.

Treating viral diseases

- Antibiotics *do not* affect viruses.
- Drugs that kill viruses often damage the body's tissues.
- Painkillers treat the symptoms of viral diseases but do not kill pathogens.

Discovering and developing new drugs

Drugs were traditionally extracted from plants and microorganisms, for example:

- the heart drug digitalis comes from foxglove plants
- the painkiller aspirin originates from willow trees
- penicillin was discovered by Alexander Fleming from *Penicillium* mould.

Most modern drugs are now synthesised by chemists in laboratories.

New drugs are extensively tested and trialled for

- **toxicity** – is it harmful?
- **efficacy** – does it work?
- **dose** – what amount is safe and effective to give?

Vaccination

Vaccination involves injecting small quantities of dead or inactive forms of a pathogen into the body.

This stimulates lymphocytes to produce the correct antibodies for that pathogen.

If the same pathogen re-enters the body, the correct antibodies can be produced quickly to prevent infection.

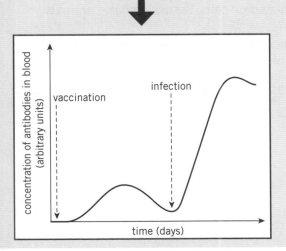

Stages of clinical trials

Pre-clinical trials

Drug is tested in cells, tissues, and live animals.

Clinical trials

1 Healthy volunteers receive very low doses to test whether the drug is safe and effective.
2 If safe, larger numbers of healthy volunteers and patients receive the drug to find the optimum dose.

Peer review

Before being published, the results of clinical trials will be tested and checked by independent researchers. This is called peer review.

Double-blind trials

Some clinical trials give some of their patients a **placebo** drug – one that is known to have no effect.

Double-blind trials are when neither the patients nor the doctors know who has been given the real drug and who has been given the placebo. This reduces biases in the trial.

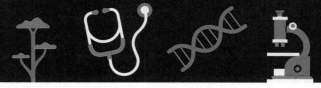

Learn the answers to the questions below then cover the answers column with a piece of paper and write as many as you can. Check and repeat.

B8 questions	Answers
1 What non-specific systems does the body use to prevent pathogens getting into it?	skin; cilia and mucus in the nose, trachea, and bronchi; stomach acid
2 What three functions do white blood cells have?	phagocytosis, producing antibodies, producing antitoxins
3 What happens during phagocytosis?	phagocyte is attracted to the area of infection, engulfs a pathogen, and releases enzymes to digest the pathogen
4 What are antigens?	proteins on the surface of a pathogen
5 Why are antibodies a specific defence?	antibodies have to be the right shape for a pathogen's unique antigens, so they target a specific pathogen
6 What is the function of an antitoxin?	neutralise toxins produced by pathogens by binding to them
7 What does a vaccine contain?	small quantities of a dead or inactive form of a pathogen
8 How does vaccination protect against a specific pathogen?	vaccination stimulates the body to produce antibodies against a specific pathogen – if the same pathogen re-enters the body, white blood cells rapidly produce the correct antibodies
9 What is herd immunity?	when most of a population is vaccinated against a disease, meaning it is less likely to spread
10 What is an antibiotic?	drugs that kill bacteria but not viruses
11 What do painkillers do?	treat some symptoms of diseases and relieve pain
12 What properties of new drugs are clinical trials designed to test?	toxicity, efficacy, and optimum dose
13 What happens in the pre-clinical stage of a drug trial?	drug is tested on cells, tissues, and live animals
14 What is a placebo?	medicine with no effect that is given to patients instead of the real drug in a trial
15 What is a double-blind trial?	a trial where neither patients nor doctors know who receives the real drug and who receives the placebo

Put paper here

Now go back and use the questions below to check your knowledge from previous chapters.

Previous questions

Answers

Previous questions		Answers
What is an organ?	Put paper here	group of tissues working together to perform a specific function
Why is a leaf an organ?		there are many tissues inside the leaf that work together to perform photosynthesis
How do white blood cells protect the body?	Put paper here	• engulf pathogens • produce antitoxins to neutralise toxins, or antibodies
How can plant diseases be detected?	Put paper here	areas of decay, discolouration, growths, malformed stems or leaves, presence of pests, spots on leaves, and stunted growth
Name four factors that affect the rate of transpiration.	Put paper here	temperature, light intensity, humidity, and wind speed
What are three mechanical defences that protect plants?		thorns and hairs, leaves that droop or curl, mimicry to trick animals
What happens during the third stage of the cell cycle?		the cytoplasm and cell membrane divide, forming two identical daughter cells

Maths Skills

Practise your maths skills using the worked example and practice questions below.

Standard form

Standard form is a way of writing very large or very small numbers. For example, in biology, we can use standard form when working with the size of cells and organelles as they are so small.

When writing a number in standard form, you first write a digit between 1 and 10, then you write $\times 10^n$, where the power of ten expresses how big or small the number is.

For large numbers, positive powers of ten shift the digit to the left:

$23\,000\,000 = 2.3 \times 10^7$

For small numbers, negative powers of ten shift the digit to the right:

$0.000\,000\,23 = 2.3 \times 10^{-7}$

Worked Example

Examples of powers of 10:

10^1	10
10^3	1000
10^7	10 000 000
10^{-3}	0.001
10^{-7}	0.000 000 1

What is 70 000 0 written in standard form?

70 000 0 can be written as
$7 \times 10\,0000$

$10\,0000 = 10 \times 10 \times 10 \times 10 \times 10 = 10^5$

so $70\,0000 = 7 \times 10^5$

What is 0.000 4 written in standard form?

0.000 4 can be written as $4 \times 0.000\,1$

$0.000\,1 = 10^{-4}$

So $0.000\,4 = 4 \times 10^{-4}$

Practice

1. The World Health Organisation estimates that 3×10^8 people are infected with malaria every year. Convert this number to an expanded figure.

2. Scientists estimate that malaria kills 2×10^6 people every year. Convert this number to an expanded figure.

3. The table below gives data relating to diabetes in the UK. Write the figures in the table in standard form.

	Figure	Standard Form
population of UK in 2015	65 000 000	
number of people diagnosed with diabetes	34 500 00	
estimated number of people with undiagnosed diabetes	54 900 0	

Exam-style questions

01 Most children in the UK are vaccinated against tetanus.

Tetanus is a serious disease caused by a bacterial toxin that affects the nervous system.

01.1 Name the component in the vaccine that will make a child immune to tetanus. **[1 mark]**

_____

01.2 A few weeks after vaccination, a child becomes infected with the bacteria that cause tetanus.

Figure 1 shows the number of tetanus antibodies present in the child's blood.

Use your knowledge of vaccination to complete the graph to show what you think will happen to the number of tetanus antibodies present in the child's blood. **[2 marks]**

Figure 1

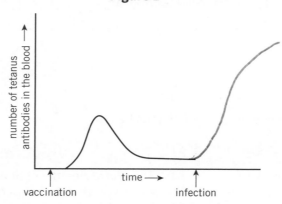

> **Exam Tip**
>
> Sometimes the scale of the graph can give you a clue.

01.3 Alongside tetanus, mumps is another disease for which vaccination is available.

Explain the advantages of vaccinating a large proportion of the population against mumps. **[2 marks]**

> **Exam Tip**
>
> Don't waste time writing down anything that isn't an advantage.

_____

_____

_____

01.4 Another vaccination offers protection against the bacteria that cause some strains of meningitis.

Explain why a person who has only received vaccinations against tetanus and mumps would not have protection against meningitis. **[3 marks]**

_____

_____

_____

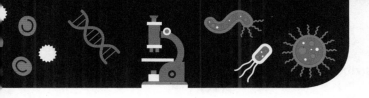
02   **Figure 2** shows a section of nasal epithelium (tissue from the nose).

Figure 2

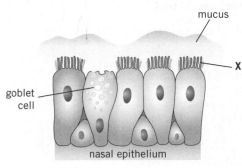

02.1   Explain why this is an example of a tissue.   **[1 mark]**

_____

_____

02.2   Suggest the function of the goblet cells.   **[1 mark]**

_____

> **! Exam Tip**
>
> Think about the location and the name of the cell.

02.3   Identify structure **X** in **Figure 2** and explain its role in defence against disease.   **[3 marks]**

_____

_____

_____

02.4   The skin is another example of a non-specific defence against disease.
It acts as a barrier to microorganism entry.
Explain how the skin protects itself after its surface is punctured.   **[3 marks]**

_____

_____

_____

02.5   Describe **one** other way the skin protects the body from disease.   **[2 marks]**

_____

_____

> **! Exam Tip**
>
> This question is just asking for you to say what happens, not why.

03 Before a medical drug can be licensed to be prescribed by a doctor, it has to undergo a number of stages of testing.

03.1 Draw **one** line from each stage of drug testing, to its purpose. **[3 marks]**

① Exam Tip

Only draw one line from each box on the left.

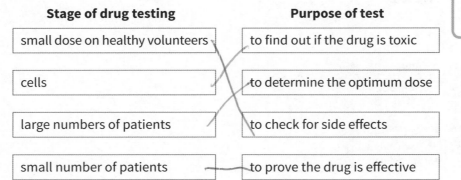

Stage of drug testing	Purpose of test
small dose on healthy volunteers	to find out if the drug is toxic
cells	to determine the optimum dose
large numbers of patients	to check for side effects
small number of patients	to prove the drug is effective

03.2 Drug trials are very expensive. They are often paid for by the company who wishes to manufacture the drug. Suggest **one** reason why people who are employed by the manufacturing company should not take part in the trial. **[1 mark]**

① Exam Tip

This may sound like a question in a business exam, but it's relevant for science as well.

03.3 Describe what is meant by a double blind trial. **[3 marks]**

03.4 Describe **one** reason why the placebo drug in a double blind trial should contain an existing treatment for the condition being targeted. **[1 mark]**

04 Since the discovery of penicillin as an antibiotic drug, many other antibiotics have been created.

04.1 Describe how penicillin was discovered. **[3 marks]**

04.2 Explain why an antibiotic drug would not be prescribed to treat a measles infection. **[2 marks]**

① Exam Tip

The social history of science is just as important as the results.

04.3 Penicillin and erythromycin are two types of antibiotic drugs. Explain why doctors have a range of antibiotic drugs available to prescribe. **[2 marks]**

04.4 Erythromycin is taken orally in tablet form. It is possible to cover the tablet in a coating that affects the rate at which the drug is absorbed into the bloodstream. **Figure 3** shows the level of erythromycin in the bloodstream over time for both forms of the tablet – coated and uncoated. Erythromycin is taken every 12 hours.

Figure 3

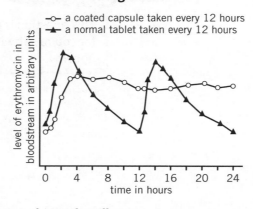

① Exam Tip

For this question, it's important you describe both lines carefully, use data from the graph, and talk about any changes that you can see over time.

Using **Figure 3**, compare the effects of the two types of tablet on a patient suffering from a bacterial infection. **[6 marks]**

05 Before medical drugs can be used on patients, they undergo a number of stages of testing.

05.1 One important test is for drug efficacy. Describe what is meant by this term. **[1 mark]**

05.2 Before clinical trials, drugs are tested in the laboratory. During laboratory trials the drugs are not tested on people. Name **two** ways the drug could be tested during this stage of the trials. **[2 marks]**

05.3 Clinical trials consist of several stages of testing. During the first phase (Phase 1) of clinical trials the potential new drug is tested on healthy volunteers. Describe what happens during the next phase (Phase 2) of testing. **[2 marks]**

! **Exam Tip**

Showing your working is a must, as you might get some marks even if you get the answer wrong.

05.4 Phase 2 clinical trials are where most drugs fail. Only 31% of drugs that enter Phase 2 studies go on to Phase 3. A pharmaceuticals company develops 13 compounds that are successful after Phase 1 of clinical trials. Estimate how many compounds the company will advance to Phase 3. **[2 marks]**

05.5 Cancer drugs have the lowest overall rate of success during clinical trials. Only 5.1% of the drugs that enter Phase 1 are ultimately approved. In 2018, 59 new cancer drugs were approved for use in patients. Estimate the number of drug compounds that were successfully tested in the laboratory to produce this number of new treatments. **[3 marks]**

06 Measles used to be a very common disease in children. It spreads quickly through groups of people through droplet infection.

06.1 Explain why measles cannot be treated using antibiotics. **[1 mark]**

06.2 People who have been infected with measles may be advised to take aspirin. Identify the plant that aspirin originates from. **[1 mark]**

foxglove   willow tree   mould   tomatoes

! **Exam Tip**

You'll get no marks if you choose more than one answer.

06.3 Explain why aspirin may be beneficial for a person with measles, even though it does not cure the disease. **[2 marks]**

06.4 The best way to prevent the spread of measles is through a national vaccination programme. Describe how the measles vaccine works. **[4 marks]**

! **Exam Tip**

This question isn't asking about *why* the levels of vaccination have dropped, but what problems this has caused.

06.5 In 2017–2018, 91.2% of children in the UK were vaccinated against measles. This was the lowest recorded vaccination level since 2010–2011. Suggest how this will affect the number of people who are infected with measles. **[2 marks]**

07 Diphtheria is a highly contagious condition which affects the nose, throat, and skin. It is transmitted through droplet infection, or by sharing items such as cups or cutlery with an infected person. **Figure 4** shows the number of diphtheria cases in England and Wales, and the number of deaths due to the illness, between 1914 and 2014.

Figure 4

07.1 Identify the highest annual number of deaths due to diphtheria. **[1 mark]**

Exam Tip

Draw lines on the graph to help you work this out.

07.2 Suggest **two** reasons why the year with the highest number of deaths due to diphtheria does not correspond to the year with the highest number of cases of diphtheria. **[2 marks]**

Exam Tip

Don't just think about the number of cases, but also what happens when people get ill.

07.3 Calculate the percentage change in the number of cases of diphtheria between 1914 and 1944. **[2 marks]**

Exam Tip

Use data from the graph.

07.4 Using **Figure 4**, suggest and explain which year the diphtheria vaccine was introduced in England and Wales. **[3 marks]**

07.5 Due to the effective childhood vaccination programme, diphtheria is an extremely rare disease in England and Wales. However, there were 14 confirmed cases of diphtheria in 2014. Suggest why diphtheria has not been eliminated entirely as a disease from England and Wales. **[3 marks]**

Exam Tip

This can be a very emotive issue – stick to the facts in a science exam.

08 The human body uses a number of defence mechanisms to protect itself against disease.

08.1 Identify which of the following is a non-specific defence mechanism. Choose **one** answer. **[1 mark]**

antibody production

antitoxin production

presence of hydrochloric acid in the stomach

production of bile

08.2 A student reads that 'a link exists between a person's white blood cell count and their risk of developing a communicable disease'. Explain why this statement is true. **[6 marks]**

Exam Tip

You must refer to the question in your answer.

09 An increasing number of strains of bacteria are resistant to all known antibiotics. Scientists are working to develop new forms of antibacterial drugs to meet this challenge. Describe the main steps scientists follow to develop a new medicine, so that it is available to be prescribed by a doctor. **[6 marks]**

10 Fatty material can build up inside our arteries. This prevents blood flow and can lead to cardiovascular disease (CVD).

10.1 Explain whether this is an example of a communicable disease. **[2 marks]**

10.2 A clinical trial was carried out to investigate the effect of a drug on the mass of the fatty deposits in patients with CVD. Explain what is meant by a double blind trial. **[2 marks]**

10.3 The doctors measured the mass of the fatty deposits before and after treatment with the drug. The mean change in mass of the fatty acid build-up in patients was calculated. The results are shown in **Table 1**.

Table 1

Group	Mean change in mass of fatty deposit in mg	Uncertainty in change in mass in mg
placebo	+10	±20
treatment	−50	±50

> **! Exam Tip**
>
> ±20 means that it can be 20 above or below the value.

Describe the results of this trial. **[4 marks]**

10.4 Evaluate the success of the drug as a treatment for CVD. **[4 marks]**

11 A chef prepares a dessert. They cut up some strawberries, put them in a bowl, and add sugar. The bowl is then left at room temperature (**Figure 5**). A few hours later, the fruit is surrounded by syrup. Syrup is a concentrated sugar solution.

Figure 5

pieces of strawberry   sugar   syrup

dessert when prepared   dessert after several hours

11.1 Explain why the syrup formed around the strawberries. **[2 marks]**

> **! Exam Tip**
>
> Look at the size of the strawberries in **Figure 5**.

11.2 Explain how the volume of syrup would have been different if the chef had left the bowl of strawberries and sugar in the fridge. **[3 marks]**

> **! Exam Tip**
>
> This links to chemistry. Think about how quickly things happen at different temperatures.

11.3 The chef investigates how the size of the strawberry pieces affects the time taken to produce $10\,cm^3$ of syrup. The results are in **Table 2**.

Table 2

Surface area of strawberry pieces in cm²	Time taken in min
8	240
10	190
16	120
18	105
20	95

Plot the data from **Table 2** onto **Figure 6**. **[2 marks]**

11.4 Draw a line of best fit on **Figure 6**. **[1 mark]**

Figure 6

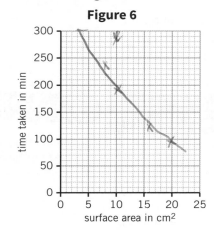

Exam Tip

Use crosses to plot points, and remember to draw a line of best fit.

11.5 Identify and explain the trend shown by the graph. **[3 marks]**

12 **Figure 7** shows actively dividing cells from the root tip of a squash plant.

Figure 7

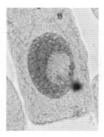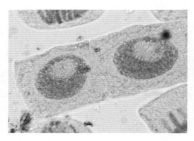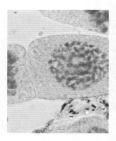

A   B   C   D   E

12.1 Name the process by which cells in a root tip divide. **[1 mark]**

12.2 Choose the correct order for the stages of division shown. **[1 mark]**

A → B → E → D → C

C → E → A → B → D ✓

C → E → B → A → D

D → B → A → E → C

13 A student is given samples of two different types of plant fluid. One was collected from a phloem vessel and one from a xylem vessel.

13.1 Describe the structure of the phloem vessel. **[2 marks]**

13.2 The student is asked to work out which plant vessel each sample has been taken from. To do this the student measures the pH, and the sugar and nitrate ion concentration, of each fluid. The results are shown in **Table 3**.

Table 3

	Sample A	Sample B
pH	7.2	5.8
sugar concentration in mg/cm³	115	0.8
nitrate ion concentration in mg/cm³	250	490

Use **Table 3** and your own knowledge to identify which sample was taken from the phloem vessel. Give reasons for your answer. **[2 marks]**

Exam Tip

You must give examples from the table in your answer.

13.3 Calculate how many times greater the concentration of nitrate ions is in sample **B** compared to sample **A**. Give your answer to two significant figures. **[2 marks]**

13.4 Maple syrup has a sugar concentration of 1.38 g/cm³. Calculate how much more concentrated the sugar content of maple syrup is compared to sample **A**. **[3 marks]**

14 A student measured how fast their heart was beating. They counted 17 beats in 15 seconds.

14.1 Calculate the student's resting heart rate in beats per minute. **[1 mark]**

14.2 The natural resting rate of the heart is maintained by a group of cells which act as a pacemaker. Identify where these cells are found. **[1 mark]**

left atrium   right atrium   left ventricle   right ventricle

14.3 Describe how an artificial pacemaker works. **[3 marks]**

Knowledge

B9 Monoclonal antibodies

Producing monoclonal antibodies

Monoclonal antibodies are produced from a single **clone** of cells.

1 Mice are injected to stimulate the production of **lymphocytes** that make specific antibodies.

Lymphocytes make antibodies but *cannot* divide to form clones

2 Tumour cells are cultured. These cells can divide and grow endlessly.

Tumour cells *can* divide to form clones

3 The lymphocytes are fused with the tumour cells to create **hybridoma** cells.

A single hybridoma cell can divide to make a large number of identical cells called a clone.

All the cloned cells can make the antibody.

4 A large amount of the monoclonal antibody can then be produced, collected, and purified for use.

Use of monoclonal antibodies

Monoclonal antibodies are specific to a single binding site on a specific protein antigen.

This means they can be used to target specific chemicals or cells.

Research

Specific molecules can be located in cells and tissues by using monoclonal antibodies to bind them to a fluorescent dye.

Treatment

Monoclonal antibodies can deliver toxic chemicals and drugs specifically to cancer cells, limiting their harm to other cells in the body.

 **Revision tip**

Lymphocytes protect the host by producing antibodies with many variations to bind a single antigen or pathogen in different ways. As the name suggests, a monoclonal antibody is a single variant for which all of the molecules are identical and bind in the same, predictable way.

Diagnostic testing

Monoclonal antibodies can be used to measure the levels of a particular chemical in the blood or to detect pathogens.

Pregnancy tests

Pregnant women produce the hormone **HCG**, which is excreted in their urine.

Monoclonal antibodies can be used to detect HCG in a pregnant woman's urine:

1 Urine is applied to the end of the stick.

2 The test stick contains monoclonal antibodies that only bind HCG, attached to a dye.

3 If HCG is present in the urine, the monoclonal antibodies cause a line of dye to appear. This means the pregnancy test is positive.

4 A second line appears in the control zone to show the test is valid, even if the result is negative.

Monoclonal antibodies produce more side effects than researchers expected when they were first developed.

As a result, they are not yet as widely used as was hoped.

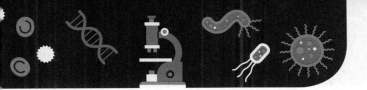

Culturing microorganisms

Bacteria multiply by simple cell division (**binary fission**).

If they have enough nutrients and the optimum temperature, the number of bacteria can double every 20 minutes.

Bacteria can be grown in a solution (**nutrient broth**) or as colonies on an **agar gel plate**.

Good **aseptic** technique is important for growing uncontaminated cultures of bacteria.

Step	Good aseptic technique and safety	Purpose
1 before culturing	**sterilise** culture media and agar before use (kill any unwanted microorganisms)	ensure no **contamination** in the media
	wipe bench/table with disinfectant	kill any microorganisms on the surface
	pass inoculating loop through a blue Bunsen flame and allow to cool slightly	sterilise the inoculating loop
2 loosen the lid on the bacteria culture bottle and dip the inoculating loop in the culture	lift the lid as little as possible when dipping the inoculating loop	reduce the chance of contamination of the culture by microorganisms from the air
3 **inoculate** the agar gel plate by streaking the inoculating loop across the surface of the agar	lift one side of the agar gel plate's lid as little as possible when inoculating the agar	reduce the chance of contamination of the agar by microorganisms from the air
4 tape the lid shut and place the inoculated agar gel plate in an incubator	do not create an airtight seal when taping the lid on	prevent anaerobic pathogens growing
	do not incubate at temperatures higher than 25 °C	reduce growth of human pathogens whose optimum temperature is human body temperature (37 °C)
	incubate plates with the surface of the agar facing downwards	stop condensation from dripping on the agar and spreading contamination
5 after culturing	pass inoculating loop through a blue Bunsen flame and place on a heatproof mat to cool	sterilise the inoculating loop
	wipe bench/table with disinfectant	kill any microorganisms on the surface

 Revision tip

Microorganisms are cultured for a variety of reasons, including scientific research and biotechnology applications.

The cultures they are grown in can also range dramatically in volume, from small 5 ml flasks to large industrial fermenters containing several thousand litres.

Key terms

Make sure you can write a definition for these key terms.

agar gel plate   aseptic   binary fission   clone   contamination   HCG   hybridoma
inoculate   lymphocyte   monoclonal   nutrient broth   sterilise

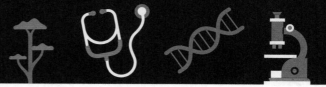
Learn the answers to the questions below then cover the answers column with a piece of paper and write as many as you can. Check and repeat.

B9 questions	Answers
1 What is a clone of cells?	group of identical cells that have formed from a single cell dividing over and over again
2 What is a hybridoma?	hybrid of a lymphocyte and tumour cell – can divide and grow endlessly, and produce antibodies
3 How are monoclonal antibodies used in research?	for locating and identifying specific molecules in cells and tissues
4 How are monoclonal antibodies used in diagnostic testing?	for measuring levels of specific hormones or chemicals in the blood or urine, for example, pregnancy tests detect HCG hormone in the urine
5 How are monoclonal antibodies used to treat cancer?	for delivering toxic chemicals and drugs directly to cancer cells, limiting their harm to other cells in the body
6 Why are monoclonal antibodies not yet as widely used as was hoped?	more side effects than were initially expected
7 Name the process by which bacteria divide.	binary fission
8 Why should an inoculating loop be passed through a blue Bunsen flame before and after use?	sterilise it/kill any bacteria
9 Name two culture media that microorganisms can be grown in.	nutrient broth solution, agar gel plates
10 Why should the lids of agar gel plates and culture bottles be opened as little as possible?	to prevent contamination with microorganisms from the air
11 Why should you not incubate at temperatures higher than 25 °C?	to reduce the chance of human pathogens growing
12 Why should agar gel plates be incubated upside down?	to prevent contamination from condensation collecting on the surface of the agar
13 How quickly can bacteria multiply?	number of bacteria can double every 20 minutes, in optimum conditions
14 Why is good aseptic technique important?	to grow bacterial cultures without contamination

Put paper here

Now go back and use the questions below to check your knowledge from previous chapters.

B9

Previous questions

Answers

Previous questions		Answers
What happens during phagocytosis?	Put paper here	phagocyte is attracted to the area of infection, engulfs a pathogen, and releases enzymes to digest the pathogen
Give three adaptations of the xylem.	Put paper here	• made of dead cells • no end wall between cells • walls strengthened by a chemical called lignin to withstand the pressure of the water
What happens during the first stage of the cell cycle?	Put paper here	cell grows bigger; chromosomes duplicate; number of subcellular structures (e.g., ribosomes and mitochondria) increases
How are fish gills adapted for efficient gas exchange?	Put paper here	• large surface area for gases to diffuse across • thin layer of cells – short diffusion pathway • good blood supply – maintains a steep concentration gradient
What is the function of the liver in digestion?		produces bile, which neutralises hydrochloric acid from the stomach and emulsifies fat to form small droplets with a large surface area

Required Practical Skills

Practise answering questions on the required practicals using the example below. You need to be able to apply your skills and knowledge to other practicals too.

Culturing microorganisms	Worked example	Practice
This practical tests your ability to observe biological changes and responses to environmental factors, in particular the effect of antiseptics and antibiotics on bacterial growth. You should be familiar with measuring, comparing, and explaining zones of inhibition of bacterial growth. You also need to be able to describe and explain aseptic technique for culturing microorganisms.	A student wanted to test how an antiseptic inhibited the growth of bacteria. They soaked three small discs in the antiseptic and placed them on a lawn of bacteria grown on an agar plate. Two days later they measured the diameter of the clear zones around the discs. The three discs had circular clear zones of 17, 19, and 20 mm in diameter. Calculate the average area of the clear zone for the antiseptic. **Answer:** average diameter $= \frac{(17 + 19 + 20)}{3} = 18.67$ mm area of circle $= \pi r^2$ $r = \frac{18.67}{2} = 9.3$ area of clear zone $= \pi \times 9.3^2 = 274$ mm^2	1 Suggest why agar plates must not be airtight when incubating. 2 Describe and explain a modification to this experiment that would make the results more valid. 3 Suggest why the discs should be spread evenly around the plate and not positioned next to each other.

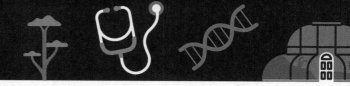

Practice

Exam-style questions

01 *Streptococcus pyogenes* is a bacterium that can cause an infection of the upper respiratory tract. This is an example of a communicable disease.

01.1 Describe what is meant by a communicable disease. **[1 mark]**

_____

_____

01.2 Scientists tested the ability of two antiseptics to kill *S. pyogenes* bacteria. They spread the bacteria on two agar plates. They then placed a small disc of filter paper, which had been soaked in antiseptic, on the centre of each dish.

Both agar plates were incubated for 24 hours.

Give **two** other variables the scientists would need to control during the test. **[2 marks]**

1 _____

2 _____

Figure 1 shows the results of their investigation after 24 hours of incubation.

Figure 1

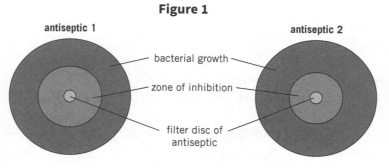

01.3 Give **two** conclusions the scientists can draw from their investigation. **[2 marks]**

1 _____

_____

2 _____

_____

01.4 Calculate the area of the zone of inhibition for antiseptic **2**. Give your answer to three significant figures. **[3 marks]**

> **! Exam Tip**
> Don't be confused by the colours in **Figure 1**.

> **! Exam Tip**
> You have to make two different measurements for this question.

Area _____ mm²

01.5 Suggest **one** reason why the scientists may not be able to recommend the use of antiseptic **1** in the home. **[1 mark]**

_____

_____

02 Monoclonal antibodies are used in pregnancy tests. They bind to the hormone human chorionic gonadotrophin (HCG). HCG is made in the early stages of pregnancy. Tiny amounts of this hormone are contained in the urine of a pregnant woman.

02.1 Explain why the monoclonal antibodies in a pregnancy test will only bind to HCG hormone. **[2 marks]**

_____

_____

_____

02.2 **Figure 2** shows the results of four pregnancy tests, each taken by a different woman.

Figure 2

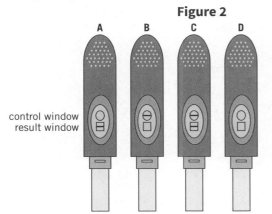

control window
result window

positive test result
a line appears in the control window and the result window.

negative test result
a line appears only in the control window.

invalid test result
no line appears in the control window.

Identify which of the women are pregnant. **[1 mark]**

_____

02.3 **Figure 3** shows how the pregnancy test is made up. HCG has more than one antibody binding site so it can bind to more than one antibody.

Figure 3

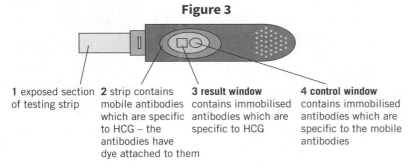

1 exposed section of testing strip

2 strip contains mobile antibodies which are specific to HCG – the antibodies have dye attached to them

3 result window contains immobilised antibodies which are specific to HCG

4 control window contains immobilised antibodies which are specific to the mobile antibodies

Explain how the pregnancy test works to show a positive result. **[6 marks]**

_____

_____

_____

_____

_____

_____

_____

Exam Tip

Even if you haven't looked at this example in class, you need to be able to apply your knowledge to this new situation. Lots of the information you need is in **Figure 3**.

03 Under optimal conditions, bacteria replicate very quickly.

03.1 Name the process by which bacteria divide. **[1 mark]**

03.2 Describe the conditions for the optimal growth of aerobic species of bacteria. **[2 marks]**

03.3 Compare the processes of bacterial division and mitosis. **[3 marks]**

Exam Tip

'Compare' means you need to discuss things that are the same and things that are different.

03.4 _Bacillus subtilis_ is a type of bacteria that is responsible for some forms of food poisoning. It divides once every 30 minutes in optimal conditions.

Calculate how many bacteria will be present after 40 hours, under optimal conditions, from a starting point of one _B. subtilis_ bacterium.

Give your answer in standard form. **[3 marks]**

04 Monoclonal antibodies are present in test strips that are used to check for the presence of malaria.

A diagram of a test strip can be seen in **Figure 4**.

Figure 4

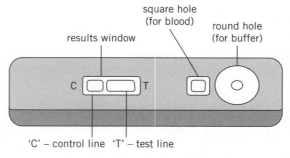

'C' – control line   'T' – test line

The strips contain monoclonal antibody **X**, which is specific to the malarial antigen. This monoclonal antibody is labelled with a dye that becomes visible when concentrated. A drop of the patient's blood is placed on the strip and a buffer solution is added. The buffer solution carries the blood and monoclonal antibody **X** along the test strip.

Exam Tip

Read the text carefully.

Highlight all the times a mobile antibody is mentioned.

The results window contains two further types of monoclonal antibody. The control line contains monoclonal antibody **Y**, which is specific to monoclonal antibody **X**. The test line contains monoclonal antibody **Z**, which is specific to the monoclonal antibody **X**–malarial antigen complex.

04.1 Identify which of the monoclonal antibodies, **X**, **Y**, or **Z**, are mobile. **[1 mark]**

04.2 A patient with malaria uses the test strip. Use the information shown above, and your own knowledge, to explain how the strip shows a positive result in the results window. **[4 marks]**

04.3 Explain the purpose of the control line. **[3 marks]**

> **! Exam Tip**
> This test works in a very similar way to a pregnancy test.

05 A student is investigating the effects of a range of household chemicals on the growth of bacteria.

05.1 Describe **two** safety precautions that they should take. **[2 marks]**

05.2 Before they could test the different chemicals, the student needed to prepare an agar plate containing an uncontaminated culture of harmless bacteria.

Describe the main steps the student should have followed to prepare this plate. **[4 marks]**

> **! Exam Tip**
> Aseptic technique is a key skill in biology.

05.3 The student then soaked a small disc of filter paper in each chemical and placed the discs on the agar plate. The student also added one disc of filter paper that had been soaked in distilled water. The agar plate was then left for one week.

The student's results are shown in **Figure 5**.

Figure 5

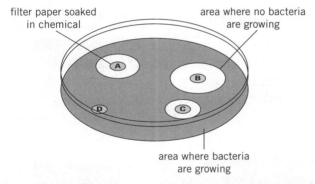

filter paper soaked in chemical — area where no bacteria are growing

area where bacteria are growing

Identify which of the discs, **A**–**D**, was soaked in distilled water. Give a reason for your answer. **[2 marks]**

05.4 Calculate how much more effective chemical **B** was than chemical **C**. The diameter of **B** = 10.4 mm, and the diameter of **C** = 6.2 mm. **[3 marks]**

> **! Exam Tip**
> The first thing you need to do is work out the area.

06 Monoclonal antibodies are antibodies that are specific to a single binding site on one protein antigen.

06.1 Describe **one** medical use of monoclonal antibodies. **[1 mark]**

06.2 Describe the process of making monoclonal antibodies. **[6 marks]**

07 *Staphylococcus aureus* is a type of bacteria commonly found on the skin and in the nose and throat of people and animals. The mean division time of *S. aureus* in optimum conditions is 30 minutes.

07.1 Calculate how long it will take for one *S. aureus* bacterium to produce a population of 256 cells. Give your answer in hours. **[3 marks]**

07.2 In less favourable conditions a *S. aureus* bacterium divides to produce 32 cells in 200 minutes. Calculate the mean division time in these conditions. **[2 marks]**

08 Scientists need to work aseptically when working with bacteria.

08.1 Identify the **two** examples of aseptic technique. **[2 marks]**

using an incubator   heating an inoculating loop until red-hot

using an autoclave   heating solutions in a water bath

08.2 Describe the main steps to produce an uncontaminated culture of bacteria on an agar plate. **[4 marks]**

08.3 After preparing an agar plate it will often be placed in an incubator to promote the growth of the bacteria. Identify the best of the conditions listed in **Table 1** (**A–E**) in which to incubate the plate. **[1 mark]**

Exam Tip

Aseptic means no bacteria will be transferred.

Exam Tip

This part of the required practical may have been done for you at school, but you still need to know the technique.

Exam Tip

Cross out any conditions that you know can lead to growth of bacteria.

Table 1

Conditions	20 °C	30 °C	Upside down	Fully sealed	Lid secured
A	✓				✓
B		✓	✓	✓	
C		✓		✓	
D	✓		✓		✓
E		✓	✓		✓

08.4 Give **two** reasons for your answer. **[2 marks]**

09 A range of chemicals can be used to kill bacteria.

09.1 Draw **one** line between each chemical and its use. **[2 marks]**

Chemical	Use
antibiotic	kill bacteria on the skin
antiseptic	kill bacteria on surfaces in the home
disinfectant	kill bacteria inside the body

09.2 A group of students tested the effectiveness of four different disinfectants on bacteria. They placed filter paper discs soaked in disinfectant onto an agar plate containing bacteria. The plates were then incubated for five days.

The results are shown in **Table 2**.

Table 2

Disinfectant	Area of clear zone on agar plate in mm²
A	2.5
B	3.6
C	2.8
D	1.8

Describe what is meant by the clear zone. **[1 mark]**

09.3 Identify which is the most effective disinfectant. **[1 mark]**

09.4 Explain how the results show that all of the solutions tested acted as a disinfectant. **[2 marks]**

10 A gardener noticed that some of the plants in their greenhouse had not grown properly. One of the symptoms of plant disease is stunted growth.

10.1 Suggest **one** other way the gardener could identify a plant with a disease. **[1 mark]**

10.2 Give **two** ways the gardener could identify the organism that had caused the disease. **[2 marks]**

10.3 Plant growth is also affected if plants do not have access to sufficient minerals. Plants lacking in magnesium have yellow leaves.

Explain how a magnesium deficiency affects plant growth. **[3 marks]**

Coronary heart disease

Coronary heart disease (CHD) occurs when the coronary arteries become narrowed by the build-up of layers of fatty material within them.

This reduces the flow of blood, resulting in less oxygen for the heart muscle, which can lead to heart attacks.

Health issues

Health is the state of physical and mental well-being.

The following factors can affect health:

- communicable and non-communicable diseases
- diet
- stress
- exercise
- life situations

Different types of disease may interact, for example

- defects in the immune system make an individual more likely to suffer from infectious diseases
- viral infection can trigger cancers
- immune reactions initially caused by a pathogen can trigger allergies, for example skin rashes and asthma
- severe physical ill health can lead to depression and other mental illnesses.

Treating cardiovascular diseases

Treatment	Description	Advantages	Disadvantages
stent	inserted into blocked coronary arteries to keep them open	• widens the artery – allows more blood to flow, so more oxygen is supplied to the heart • less serious surgery	• can involve major surgery – risk of infection, blood loss, blood clots, and damage to blood vessels • risks from anaesthetic used during surgery
statins	drugs that reduce blood **cholesterol** levels, slowing down the deposit of fatty material in the arteries	• effective • no need for surgery • can prevent CHD from developing	• possible side effects such as muscle pain, headaches, and sickness • cannot cure CHD, so patient will have to take tablets for many years
replace faulty heart valves	heart valves that leak or do not open fully, preventing control of blood flow to the heart, can be replaced with biological or mechanical valves	• allows control of blood flow to the heart • long-term cure for faulty heart valves	• can involve major surgery – risk of infection and blood loss • risks from anaesthetic used during surgery
transplants	if the heart fails a donor heart, or heart and lungs, can be transplanted **artificial hearts** can be used to keep patients alive while waiting for a heart transplant, or to allow the heart to rest during recovery	• long-term cure for the most serious heart conditions • treats problems that cannot be treated in other ways	• transplant may be rejected if there is not a match between donor and patient • lengthy process • major surgery – risk of infection and blood loss • risks from anaesthetic used during surgery

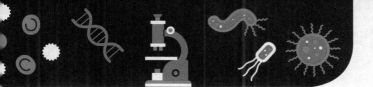

Risk factors and non-communicable diseases

A **risk factor** is any aspect of your lifestyle or substance in your body that can increase the risk of a disease developing.

Some risk factors cause specific diseases.

Other diseases are caused by factors interacting.

Risk factor	Disease	Effects of risk factor
diet (obesity) and amount of exercise	Type 2 diabetes	body does not respond properly to the production of insulin, so blood glucose levels cannot be controlled
	cardiovascular diseases	increased blood cholesterol can lead to CHD
alcohol	impaired liver function	long-term alcohol use causes liver cirrhosis (scarring), meaning the liver cannot remove toxins from the body or produce sufficient bile
	impaired brain function	damages the brain and can cause anxiety and depression
	affected development of unborn babies	alcohol can pass through the placenta, risking miscarriages, premature births, and birth defects
smoking	lung disease and cancers	cigarettes contain carcinogens which can cause cancers
	affected development of unborn babies	chemicals can pass through the placenta, risking premature births and birth defects
carcinogens, such as ionising radiation, and genetic risk factors	cancers	for example, tar in cigarettes and ultraviolet rays from the Sun can cause cancers
		some genetic factors make an individual more likely to develop certain cancers

Cancer

Cancer is the result of changes in cells that lead to uncontrolled growth and division by mitosis.

Rapid division of abnormal cells can form a **tumour**.

Malignant tumours are cancerous tumours that invade neighbouring tissues and spread to other parts of the body in the blood, forming secondary tumours.

Benign tumours are non-cancerous tumours that do not spread in the body.

Treatment of non-communicable diseases linked to lifestyle risk factors – such as poor diet, drinking alcohol, and smoking – can be very costly, both to individuals and to the Government.

A high incidence of these lifestyle risk factors can cause high rates of non-communicable diseases in a population.

 **Key terms**

Make sure you can write a definition for these key terms.

artificial heart   benign   carcinogen   cholesterol   coronary heart disease   health

malignant   risk factor   statin   stent   tumour

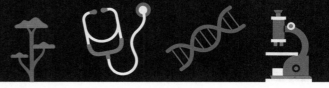
Learn the answers to the questions below then cover the answers column with a piece of paper and write as many as you can. Check and repeat.

	B10 questions		Answers
1	What is coronary heart disease?	Put paper here	layer of fatty material that builds up inside the coronary arteries, narrowing them – results in a lack of oxygen for the heart
2	What is a stent?	Put paper here	device inserted into a blocked artery to keep it open, allowing more blood and oxygen to the heart
3	What are statins?	Put paper here	drugs that reduce blood cholesterol levels, slowing the rate of fatty material deposit
4	What is a faulty heart valve?	Put paper here	valve that doesn't open properly or leaks
5	How can a faulty heart valve be treated?	Put paper here	replace with a biological or mechanical valve
6	When do heart transplants take place?	Put paper here	in cases of heart failure
7	What are artificial hearts used for?	Put paper here	keep patients alive while waiting for a transplant, or allow the heart to rest for recovery
8	Define health.	Put paper here	state of physical and mental well-being
9	What factors can affect health?	Put paper here	disease, diet, stress, exercise, life situations
10	What is a risk factor?	Put paper here	aspect of lifestyle or substance in the body that can increase the risk of a disease developing
11	Give five risk factors.	Put paper here	diet, smoking, exercise, alcohol, carcinogens
12	What is cancer?	Put paper here	a result of changes in cells that lead to uncontrolled growth and cell division by mitosis
13	What are malignant tumours?	Put paper here	cancerous tumours that can spread to neighbouring tissues and other parts of the body in the blood, forming secondary tumours
14	What are benign tumours?	Put paper here	non-cancerous tumours that do not spread in the body
15	What two types of risk factor affect the development of cancers?	Put paper here	lifestyle and genetic risk factors

Now go back and use the questions below to check your knowledge from previous chapters.

Previous questions | Answers

Previous questions		Answers
Describe an example of a fungal disease in plants.	Put paper here	rose black spot – spread by water and wind, and affects plant growth by reducing a plant's ability to photosynthesise
Describe the effect of temperature on enzyme activity.	Put paper here	as temperature increases, rate of reaction increases until it reaches the optimum for enzyme activity – above this temperature enzyme activity decreases and eventually stops
Why should the lids of agar gel plates and culture bottles be opened as little as possible?	Put paper here	to prevent contamination with microorganisms from the air
How are monoclonal antibodies used in diagnostic testing?	Put paper here	for measuring levels of specific hormones or chemicals in the blood or urine, for example, pregnancy tests detect HCG hormone in the urine
Define the term transpiration.	Put paper here	movement of water from the roots to the leaves through the xylem
What are benign tumours?		non-cancerous tumours that do not spread in the body
What is a double-blind trial?		a trial where neither patients nor doctors know who receives the real drug and who receives the placebo
What is herd immunity?		when most of a population is vaccinated against a disease, meaning it is less likely to spread

Maths Skills

Practise your maths skills using the worked example and practice questions below.

Calculating rate of blood flow	Worked Example	Practice
The rate of blood flow in the body changes in response to things like exercise and illnesses. Blood flow increases during exercise to deliver oxygen to working muscles and to remove waste products. The rate of blood flow can be reduced by non-communicable diseases such as coronary heart disease. To calculate rate of blood flow: rate of blood flow (ml/min) $= \dfrac{\text{volume of blood (ml)}}{\text{time (min)}}$ Remember to add units to your answer. Rate of blood flow can be measured in ml/min or l/hr – check the question to see which units you need to use.	1660 ml of blood is pumped through a vein in 4 min. Calculate the rate of blood flow through the vein in ml/min. $\dfrac{1660}{4} = 15\,\text{ml/min}$ You may have to convert millilitres to litres if given a large volume. To do this, divide the volume in ml by 1000.	**1** 3540 ml of blood is pumped through an artery in 3.5 min. Calculate the rate of blood flow through the artery in ml/min. **2** 11 540 ml of blood is pumped through an artery in 12.5 minutes. Calculate the rate of blood flow through the artery in ml/min. **3** 670 l of blood is pumped through the heart in 1 hr. Calculate the rate of blood flow through the heart in ml/min.

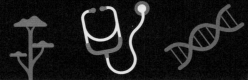

Practice

Exam-style questions

01   **Figure 1** shows the main causes of death in the UK in 2012 for people under the age of 75. The total number of deaths recorded in this period was 150 000.

Figure 1

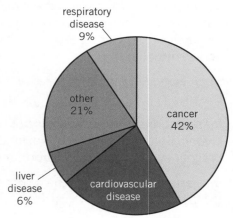

respiratory
disease
9%

other
21%

cancer
42%

liver
disease
6%

cardiovascular
disease

01.1   Describe what is meant by cardiovascular disease (CVD).   **[1 mark]**

_____

_____

01.2   Calculate the percentage of people under the age of 75 who died in 2012 due to CVD.   **[1 mark]**

_____ %

01.3   Calculate the number of people under the age of 75 who died in 2012 due to CVD.   **[2 marks]**

_____ people

Exam Tip

The total number of people is given in the first part of the question.

01.4   Explain **three** ways a person could reduce their risk of CVD.   **[6 marks]**

1 _____

_____

2 _____

_____

3 _____

_____

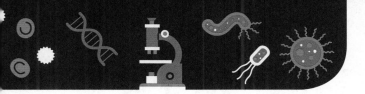

02 If a patient has a blocked blood vessel they may be treated using a stent or by undergoing bypass surgery. This is where another piece of blood vessel is used to replace the damaged vessel.

02.1 Describe how stents are used to treat blocked blood vessels. **[4 marks]**

_____

_____

_____

_____

02.2 Evaluate the use of stents to treat blocked blood vessels by explaining the risks and benefits of having a stent. **[4 marks]**

_____

_____

_____

_____

> (!) **Exam Tip**
>
> For an 'evaluate' question you have to give your opinion and then justify it. Don't worry too much about what your opinion is – the examiners are never going to quiz you on it!

02.3 **Figure 2** shows the proportion of patients suffering complications following surgery to treat cardiovascular disease (CVD). This information was collected by analysing the health records of 2500 patients, half of whom received a stent and half of whom received a bypass operation.

Figure 2

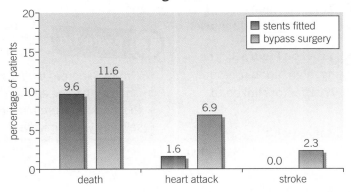

Calculate how many more patients died following a bypass operation compared to those who received a stent. **[3 marks]**

> (!) **Exam Tip**
>
> The y-axis on the graph gives you the percentage to use in this question.

 patients

03 **Figure 3** shows the effect of three key risk factors on the mortality rate of 35 000 people with cardiovascular disease (CVD) in a region of Finland. The data are all relative to the number observed in the first year of the study. In 1972

- 31 500 people had cholesterol levels above the recommended maximum level
- 11 200 people were smokers
- 28 000 people had high blood pressure
- 650 people died due to CVD.

Figure 3

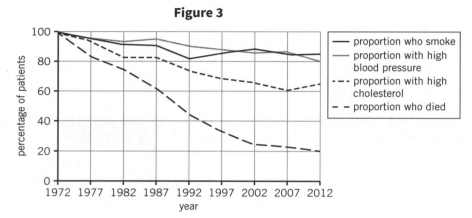

03.1 Describe the main trends shown in the three key risk factors for CVD between 1972 and 2012. **[2 marks]**

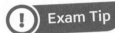

03.2 Suggest and explain **two** reasons for the trends shown in the graph. **[4 marks]**

03.3 Explain how a high level of cholesterol can cause a heart attack. **[3 marks]**

03.4 Name the drug used to reduce blood cholesterol levels. **[1 mark]**

03.5 Calculate the number of deaths due to CVD in this region of Finland in 2012. **[3 marks]**

03.6 A student writes the following conclusion based on **Figure 3**: 'The data gathered in Finland prove that smoking, high blood pressure, and high levels of cholesterol are three key risk factors linked to cardiovascular disease.'

Explain the extent to which you agree or disagree with this conclusion. **[3 marks]**

> **Exam Tip**
>
> Try highlighting the lines needed for this question in a bright colour to make them stand out and help you identify the trend.

> **Exam Tip**
>
> Writing "I agree" or "I disagree" isn't going to be enough to get the marks here, you need to say why you think that.

04 To reduce the number of deaths from non-communicable diseases, scientists study large volumes of data to search for possible links between risk factors and disease.

04.1 Describe the difference between a risk factor that shows a correlation with the incidence of a disease and a risk factor that shows causation of a disease. **[2 marks]**

04.2 Give **one** risk factor for a disease that you cannot control. **[1 mark]**

04.3 Unprotected exposure to sunlight is a risk factor for developing skin cancer.

Explain what is meant by the term cancer. **[2 marks]**

04.4 Explain how sun exposure increases a person's risk of developing skin cancer. **[2 marks]**

04.5 Smoking is a risk factor for the development of tumours. Most tumours caused by smoking are malignant.

Explain the difference between a benign and a malignant tumour. **[3 marks]**

05 Every year many patients need to have heart valve replacements.

05.1 Describe the function of the heart valve labelled **X** in **Figure 4**. **[2 marks]**

Figure 4

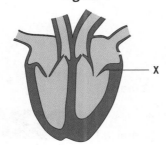

05.2 Over time valves can become leaky. Explain how this can cause health issues. **[3 marks]**

05.3 **Table 1** gives information about two types of heart valve.

Table 1

	Mechanical heart valve	Biological heart valve
material	titanium	usually cow or pig tissue, but can be from a human donor
lifespan	20 years	10–15 years
additional medication	anti-coagulation medication to prevent blood clotting around the valve	not required

> **! Exam Tip**
>
> Make sure you refer to *all* of the information given in the table – put a little tick next to it when you've used it.

A 20-year-old patient requires a heart valve replacement. Using **Table 1** and your own knowledge, evaluate the advantages and disadvantages of each type of heart valve. **[6 marks]**

06 Whilst waiting for a heart transplant some patients are fitted with artificial hearts to keep them alive.

Figure 5 shows one example of an artificial heart. It is connected to an external power supply.

Figure 5

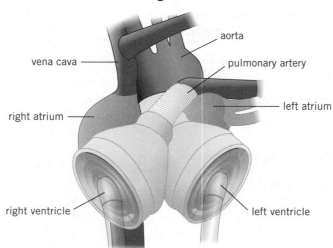

06.1 Use the information given above and your own knowledge to identify **two** differences between a real heart and an artificial heart. **[2 marks]**

06.2 Suggest **two** advantages and **two** disadvantages of treating patients with this artificial heart. **[4 marks]**

 Exam Tip

You can use a table to answer this question, making it clear what goes with what.

06.3 Suggest **one** reason why artificial hearts are not widely used in the treatment of heart disease. **[1 mark]**

07 One measure of a person's health is their body mass index (BMI), which is calculated using the following formula:

$$BMI = \frac{mass}{height^2}$$

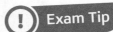

07.1 A student has a mass of 48 kg and is 1.5 metres tall. Calculate the student's BMI. **[2 marks]**

 Exam Tip

It's just the height that is squared and not everything – don't forget BODMAS.

07.2 Using **Table 2**, identify which weight category the student belongs to. **[1 mark]**

Table 2

Weight category	BMI in kg/m²
underweight	<18.5
healthy weight	18.5–24.9
overweight	25–29.9
obese	30–34.9
severely obese	35–39.9
morbidly obese	≥40

07.3 In the UK in 2016, 26% of adults were classified as obese.
Explain the effects of obesity on the body. **[6 marks]**

07.4 One public health campaign aims to increase levels of exercise amongst the population by getting people to choose to walk or cycle to work, rather than take public transport or drive.
Explain how this campaign will help to reduce obesity levels in the population. **[2 marks]**

07.5 It typically costs £2 billion to develop a new drug and get it onto the market. Evaluate the costs and benefits to society of the availability of an anti-obesity drug that suppresses a person's feeling of hunger. **[6 marks]**

08 If a person has a high level of cholesterol in their blood, it increases their risk of a heart attack or a stroke.

08.1 Explain how the build-up of cholesterol can cause a heart attack. **[4 marks]**

08.2 Give **one** factor, apart from drugs, that can affect the level of cholesterol in a person's blood. **[1 mark]**

08.3 Statins reduce blood cholesterol levels. They are commonly prescribed to people who have high levels of blood cholesterol. However, statins cause negative effects in some patients. One study showed the following information for every 10 000 people treated with statins:

- 275 fewer cases of heart disease occurred than had been predicted
- 10 fewer cases of oesophageal cancer than had been predicted
- 25 extra patients experienced acute kidney failure compared to mean levels
- 75 extra patients experienced liver dysfunction compared to mean levels
- 300 patients developed cataracts
- 150 patients experienced muscle weakness.

Explain why doctors prescribe statins despite the risks to patients of developing another medical condition. **[2 marks]**

08.4 In the year of the study, 4 200 000 people (from a population of 60 000 000) in the UK experienced liver dysfunction. Evaluate whether taking statins causes a significant increase in the risk of liver failure. **[5 marks]**

! **Exam Tip**

'Explain' questions are all about the *why*.

! **Exam Tip**

Think about ways that reducing obesity can reduce costs of healthcare.

! **Exam Tips**

Use data from the question in your answer.

This isn't testing what you know but your analysis and interpretation skills.

09 Atherosclerosis is one form of cardiovascular disease (CVD). Patients with this form of the disease have a build-up of fatty material on the inner walls of their arteries.

09.1 Explain how atherosclerosis increases the risk of a heart attack. **[2 marks]**

09.2 Describe **one** mechanical technique doctors can use to lower the risk of a heart attack for a patient with atherosclerosis. **[3 marks]**

09.3 Patients with CVD are often advised to take aspirin daily. An effect of aspirin is to reduce the ability of platelets to stick together.

Suggest and explain the benefits of a person with atherosclerosis taking this drug. **[4 marks]**

10 Fresh cow's milk is a mixture containing water, lipids, protein, and lactose sugar. It also contains some vitamins and minerals.

10.1 Describe the chemical test that could be used to show that there is protein present in milk. **[2 marks]**

10.2 Lactose cannot be absorbed into the body. It must be digested by the enzyme lactase into the sugars glucose and galactose, which can then be absorbed.

Suggest why lactose cannot be absorbed into the blood. **[1 mark]**

10.3 Lactase can be added to fresh milk to pre-digest the lactose. This makes 'lactose-free' milk, which is suitable for people who do not produce enough lactase of their own. A company that produces lactose-free milk investigated the effect of temperature on lactase. Their results are shown in **Table 3**.

Table 3

Temperature in °C	Time taken to digest lactose in min
25	20
30	14
35	11
40	11
45	29
50	no digestion

Explain why no digestion occurred at 50 °C. **[3 marks]**

10.4 Using the information provided, suggest the optimum temperature for the company to heat milk to, prior to adding the lactase enzyme. Give a reason for your answer. **[2 marks]**

11 A student looked down a microscope to observe some actively dividing cells from a root tip. They estimated that there were 700 cells in their field of view. The student counted 36 cells that were going through mitosis.

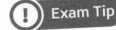

11.1 If one complete cell cycle is 25 hours, estimate how long mitosis takes. Use the equation: **[3 marks]**

length of cell cycle stage =

$$\frac{\text{observed number of cells at the stage}}{\text{observed total number of cells}} \times \text{total length of cell cycle}$$

11.2 Describe the main steps in the cell cycle. **[4 marks]**

12 A group of students set up an experiment to investigate osmosis in cells. They used two sections of Visking tubing to represent two different cells. The students filled each piece of Visking tubing with a different solution:

- cell **1** contained 20% sucrose solution
- cell **2** contained distilled water.

The students then placed both pieces of Visking tubing in a beaker containing 5% sucrose solution and left them for one hour.

The experimental set-up is shown in **Figure 6**.

Figure 6

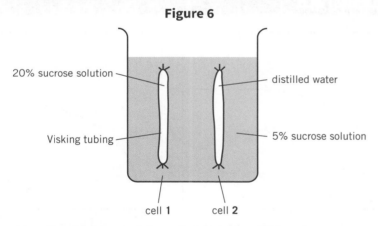

20% sucrose solution — Visking tubing — distilled water — 5% sucrose solution — cell **1** — cell **2**

12.1 Identify which part of the cell the Visking tubing represents. **[1 mark]**

12.2 Give **one** variable the students should have controlled. **[1 mark]**

12.3 Explain the results observed by the students after one hour. **[4 marks]**

Knowledge

B11 Photosynthesis

Photosynthetic reaction

Photosynthesis is a chemical reaction in which energy is transferred from the environment as light from the Sun to the leaves of a plant. This is an **endothermic** reaction.

Chlorophyll, the green pigment in **chloroplasts** in the leaves, absorbs the light energy. Leaves are well-adapted to increase rates of photosynthesis (Chapter 6 *Organisation in plants*).

Photosynthesis rate

A **limiting factor** is anything that limits the rate of a reaction when it is in short supply.

The limiting factors for photosynthesis are:
- temperature
- carbon dioxide concentration
- light intensity
- amount of chlorophyll.

Less chlorophyll in the leaves reduces the rate of photosynthesis. More chlorophyll may be produced by plants in well-lit areas to increase the photosynthesis rate.

 carbon dioxide + water $\xrightarrow{\text{light}}$ **glucose** + oxygen

$6CO_2$ + $6H_2O$ $\xrightarrow{\text{light}}$ $C_6H_{12}O_6$ + $6O_2$

Limiting factors and photosynthesis rate

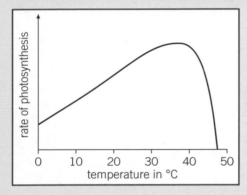

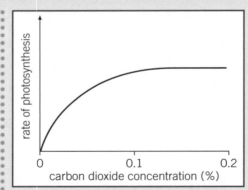

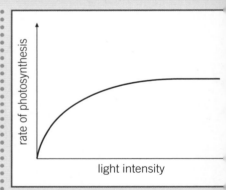

- At low temperatures the rate of photosynthesis is low because the reactant molecules have less kinetic energy.
- Photosynthesis is an enzyme-controlled reaction, so at high temperatures the enzymes are denatured and the rate quickly decreases.

- Carbon dioxide is used up in photosynthesis, so increasing carbon dioxide concentration increases the rate of photosynthesis.
- At a certain point, another factor becomes limiting.
- Carbon dioxide is often the limiting factor for photosynthesis.

- Light energy is needed for photosynthesis, so increasing light intensity increases the rate of photosynthesis.
- At a certain point, another factor becomes limiting.
- Photosynthesis will stop if there is little or no light.

 **Key terms**

Make sure you can write a definition for these key terms.

chlorophyll   chloroplast   endothermic   glucose

inverse square law   limiting factor   photosynthesis

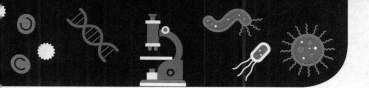

Interaction of limiting factors

Limiting factors often interact, and any one may be limiting photosynthesis.

For example, on the graph the lowest curve has both carbon dioxide and temperature limiting photosynthesis. Temperature is limiting for the middle curve, and the highest curve shows photosynthesis rate increases when both temperature and carbon dioxide are increased until another factor becomes limiting.

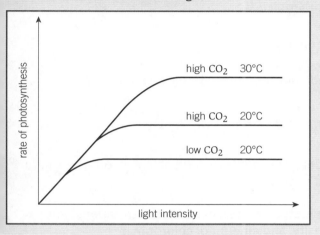

Inverse square law

As the distance of a light source from a plant increases, the light intensity decreases – an inverse relationship. This is not linear, as light intensity varies in inverse proportion to the square of the distance:

$$\text{light intensity} \propto \frac{1}{\text{distance}^2}$$

For example, if you double the distance between a light source and a plant, light intensity falls by a quarter.

Greenhouse economics

Commercial greenhouses control limiting factors to get the highest possible rates of photosynthesis so they can grow plants as quickly as possible or produce the highest yields, whilst still making profit.

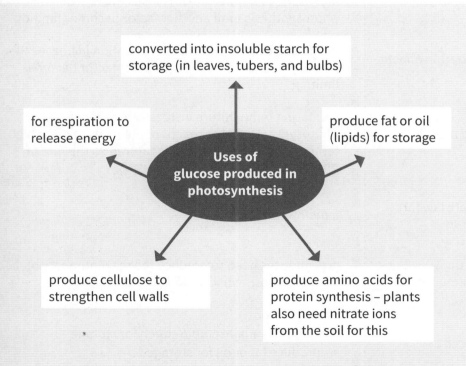

converted into insoluble starch for storage (in leaves, tubers, and bulbs)

for respiration to release energy

produce fat or oil (lipids) for storage

Uses of glucose produced in photosynthesis

produce cellulose to strengthen cell walls

produce amino acids for protein synthesis – plants also need nitrate ions from the soil for this

Revision tip

Make sure you learn the shapes of the graphs on these pages.

In an exam you may be asked to sketch them (the axes and shape of line), describe them (use words to show the shape), or explain them (say *why* the shape is how it is).

Retrieval

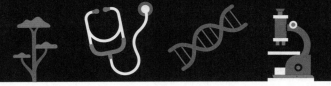

Learn the answers to the questions below then cover the answers column with a piece of paper and write as many as you can. Check and repeat.

B11 questions

Answers

#	Question	Answer
1	Where does photosynthesis occur?	chloroplasts in the leaves of a plant
2	What is the name of the green pigment in the leaves?	chlorophyll
3	What type of reaction is photosynthesis?	endothermic
4	What type of energy is used in photosynthesis?	light energy
5	Give the word equation for photosynthesis.	carbon dioxide + water $\rightarrow$ glucose + oxygen
6	Give the balanced symbol equation for photosynthesis.	$6CO_2 + 6H_2O \rightarrow C_6H_{12}O_6 + 6O_2$
7	Define the term limiting factor.	anything that limits the rate of a reaction when it is in short supply
8	Give the limiting factors of photosynthesis.	temperature, carbon dioxide concentration, light intensity, and amount of chlorophyll
9	Describe how light intensity affects the rate of photosynthesis.	increasing light intensity increases the rate of photosynthesis until another factor becomes limiting
10	Describe how carbon dioxide concentration affects the rate of photosynthesis.	increasing carbon dioxide concentration increases the rate of photosynthesis until another factor becomes limiting
11	Describe how temperature affects the rate of photosynthesis.	increasing temperature increases the rate of photosynthesis as the reaction rate increases – at high temperatures enzymes are denatured so the rate of photosynthesis quickly decreases
12	Give the equation for the inverse square law for light intensity.	light intensity $\propto \dfrac{1}{distance^2}$
13	Why are limiting factors important in the economics of growing plants in greenhouses?	greenhouses need to produce the maximum rate of photosynthesis while making profit
14	How do plants use the glucose produced in photosynthesis?	• respiration • convert it into insoluble starch for storage • produce fat or oil for storage • produce cellulose to strengthen the cell wall • produce amino acids for protein synthesis

Put paper here

Now go back and use the questions below to check your knowledge from previous chapters.

Previous questions

Answers

Previous questions		Answers
What is a hybridoma?	Put paper here	hybrid of a lymphocyte and tumour cell – can divide and grow endlessly, and produce antibodies
What three functions do white blood cells have?		phagocytosis, producing antibodies, producing antitoxins
Why is the human circulatory system a double circulatory system?	Put paper here	blood passes through the heart twice for every circuit around the body – deoxygenated blood is pumped from the right side of the heart to the lungs, and the oxygenated blood that returns is pumped from the left side of the heart to the body
Give two disadvantages of using embryonic stem cells.	Put paper here	• ethical issues surrounding their use, as every embryo is a potential life • potential risks involved with treatments, such as transfer of viral infections
Why do different digestive enzymes have different optimum pHs?	Put paper here	different parts of the digestive system have very different pHs – the stomach is strongly acidic, and the pH in the small intestine is close to neutral
What is a stent?		device inserted into a blocked artery to keep it open, allowing more blood and oxygen to the heart

 Required Practical Skills

Practise answering questions on the required practicals using the example below.
You need to be able to apply your skills and knowledge to other practicals too.

Rate of photosynthesis	Worked example	Practice
You should be able to accurately measure changes in the rate of photosynthesis of a plant, and how the rate changes in response to changes in the environment. This requires being able to describe how to measure the rate of a reaction or biological process by collecting a gas produced. For example, collecting bubbles of oxygen produced by pondweed to compare the volume of gas produced at different light intensities. It is important to understand how different factors affect rates of photosynthesis, including light intensity, temperature, and carbon dioxide concentration.	A student used an inverted test tube to investigate the number of bubbles released from a piece of pondweed in a beaker of water in a 10 minute period. They repeated each measurement 5 times. **1** Identify the dependent variable in this experiment. Number of bubbles released. **2** Explain how the student could use this set up to investigate how light intensity affects the rate of photosynthesis. Carry out the experiment described above with a switched on lamp placed exactly 10 cm from the pondweed. Record number of bubbles produced over the 10 mins, repeating experiment 5 times. Move lamp 10 cm further away from the pondweed, and repeat the same experiment. Calculate the mean number of bubbles produced for each light intensity, and compare the results.	**1** Suggest how the student could change the experiment to give a more accurate measurement of the gas released. **2** Explain how temperature affects the rate of plant photosynthesis. **3** Name a piece of equipment that could be used to investigate how temperature affects the amount of gas released by the pondweed.

Practice

Exam-style questions

01 A student set up the apparatus in **Figure 1** to investigate the effect of light intensity on photosynthesis in pondweed.

Figure 1

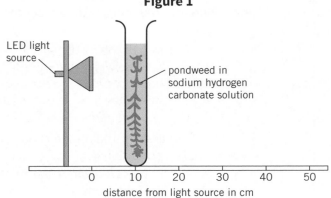

LED light source

pondweed in sodium hydrogen carbonate solution

distance from light source in cm

> **! Exam Tip**
>
> Figures can give you lots of information that you may not have thought of. For example, why is it an LED light source in **Figure 1**? Why is the pondweed not just in water?

01.1 Identify the independent variable in this investigation. **[1 mark]**

_____

01.2 Give the reason the pondweed was placed in a solution of sodium hydrogen carbonate. **[1 mark]**

_____

01.3 The student measured the rate of photosynthesis of the pondweed by counting the number of oxygen bubbles produced in one minute.
Describe how you could test the bubbles to show they contained oxygen. **[2 marks]**

_____

_____

> **! Exam Tip**
>
> You'll need a little bit of chemistry here!

01.4 The student's results are shown in **Table 1**.

Plot the data on **Figure 2**. **[4 marks]**

Table 1

Distance from light source in cm	Number of bubbles produced per minute
10	15
20	8
30	4
40	2
50	0

Figure 2

number of bubbles produced per minute

distance from light source in cm

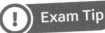

01.5   Describe the trend shown by your graph in **Figure 2**.   **[1 mark]**

_____

_____

! Exam Tip

For a one mark question you only need to give a simple answer, such as an increase or decrease.

01.6   Use **Figure 2** to determine the rate of photosynthesis at 25 cm.   **[1 mark]**

_____

01.7   The student stated that counting bubbles was not an accurate way of measuring the volume of oxygen produced.
Write down **one** reason why the student is correct.   **[1 mark]**

_____

01.8   Suggest **one** improvement the student could make to improve the accuracy of this investigation.   **[1 mark]**

_____

02   Plants produce glucose by the process of photosynthesis.

02.1   Complete the following chemical equation that describes the process of photosynthesis.   **[2 marks]**

$$6CO_2 + \underline{\hspace{2cm}} \rightarrow \underline{\hspace{2cm}} + 6O_2$$

! Exam Tip

Balancing this equation can seem tricky – the easiest thing to do is to learn the numbers.

02.2   Photosynthesis is an endothermic reaction.
Explain why this is.   **[3 marks]**

_____

_____

_____

_____

02.3   Describe how a carbon atom from the atmosphere can become part of a starch molecule inside a leaf.   **[6 marks]**

_____

_____

_____

_____

_____

_____

_____

03   **Figure 3** shows a cross-section through a leaf.

Figure 3

03.1   Identify the structures labelled **A** and **B**.   **[2 marks]**

03.2   Explain **two** ways the leaf is adapted to absorb light for photosynthesis.   **[4 marks]**

Exam Tip

You'll need to refer to specific parts of the leaf and how they act in this answer.

03.3   Explain **two** ways the leaf is adapted to take in carbon dioxide for photosynthesis.   **[4 marks]**

04   A tomato grower wants to know which growing conditions lead to the highest growth rate of tomato plants. They set up three experiments to measure the rate of photosynthesis in the tomatoes. **Figure 4** shows the results. All of the plants were given an adequate supply of water.

Figure 4

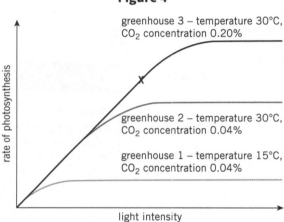

greenhouse 3 – temperature 30°C, CO_2 concentration 0.20%

greenhouse 2 – temperature 30°C, CO_2 concentration 0.04%

greenhouse 1 – temperature 15°C, CO_2 concentration 0.04%

rate of photosynthesis

light intensity

Exam Tip

Start this question by looking at the data on the graph and highlighting the differences between the conditions.

04.1   Identify the limiting factor at the point labelled **X** on the graph.   **[1 mark]**

light intensity   volume of soil added

carbon dioxide concentration   temperature

04.2   The tomato grower sets up a fourth greenhouse at a temperature of 40°C and a carbon dioxide concentration of 0.2%. Explain what will happen to the rate of photosynthesis under these conditions.   **[2 marks]**

04.3   The tomato grower concludes from their investigation that increasing temperature, light intensity, and carbon dioxide concentration maximises the rate of growth of tomato plants. Evaluate the validity of the tomato grower's conclusions.   **[3 marks]**

Exam Tip

For an 'evaluate' question you need to give a justified opinion.

04.4 Explain why the tomato grower should not raise the temperature of the greenhouses above 40 °C. **[2 marks]**

05 A number of factors affect the rate of photosynthesis. Two of these factors include carbon dioxide and temperature.

05.1 Name **one** other factor that limits the rate of photosynthesis. **[1 mark]**

05.2 **Figure 5** shows the effect of carbon dioxide on the rate of photosynthesis. Describe and explain the shape of the graph. **[4 marks]**

05.3 Explain how temperature affects the rate of photosynthesis. **[4 marks]**

Figure 5

(!) Exam Tip

This answer needs to have the *what* and the *why*.

06 Light intensity, carbon dioxide concentration, and temperature are three of the factors that affect the rate of photosynthesis. Design an investigation to study the effect of light intensity on the rate of photosynthesis. **[6 marks]**

Here is a list of some of the apparatus you might use:

- desk lamp
- pondweed
- metre rule
- beaker
- funnel

(!) Exam Tip

This is a required practical, so you should know this method really well!

07 Plants produce glucose through the process of photosynthesis. Some of this is used immediately in respiration. The remainder is converted into starch molecules.

07.1 Explain why plants convert glucose into starch. **[3 marks]**

07.2 Onions store starch in bulbs. Suggest how you could demonstrate that an onion contains starch. **[2 marks]**

07.3 Some of the glucose produced in respiration is also used to produce proteins. Explain how glucose is used to make proteins. **[4 marks]**

08 A group of students investigated the effect of temperature on the rate of photosynthesis. They carried out their investigation using pondweed. **Figure 6** shows how they set up their apparatus. The experiment was repeated using water baths set at different temperatures.

Figure 6

08.1 Write down **one** factor the students must control. **[1 mark]**

08.2 Explain why the students should leave the test tube in the water bath for five minutes before taking their measurements. **[1 mark]**

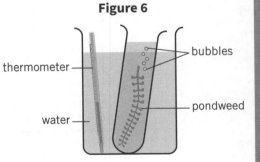

08.3 Bubbles of oxygen gas are released when the pondweed photosynthesises. Describe **one** way the students could calculate the rate of photosynthesis. **[2 marks]**

08.4 At 60 °C no bubbles of gas were produced. Explain why. **[2 marks]**

09 A scientist investigated how light intensity affected the rate of photosynthesis of pondweed. The scientist placed the pondweed in a beaker of water different distances away from a table lamp. They calculated the rate of photosynthesis by counting how many bubbles were produced in one minute. The scientist's results are shown in **Table 2**.

Table 2

Distance of lamp from pondweed in m	Light intensity in arbitrary units	Bubbles produced per minute
0.20		60
0.30	11	28
0.40	6	17
0.50	4	12
0.60	3	9

Light intensity can be calculated using the inverse square law:

$$\text{light intensity} \propto \frac{1}{\text{distance}^2}$$

> **! Exam Tip**
>
> This is the same as the inverse square law in maths.

09.1 Calculate the light intensity when the lamp is 20 cm from the pondweed. **[1 mark]**

09.2 Plot a graph of light intensity against rate of photosynthesis on **Figure 7**. **[4 marks]**

Figure 7

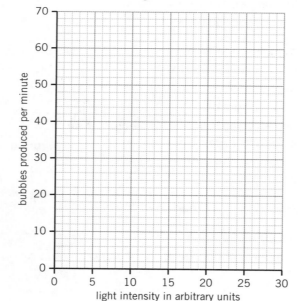

> **! Exam Tip**
>
> Always use crosses to plot points, and remember to draw a line of best fit.

> **! Exam Tip**
>
> Don't worry if you've never come across arbitrary units before, it just means it doesn't really matter what the units are.

09.3 Describe the trends shown in the data collected. **[2 marks]**

09.4 The scientist concluded that during this experiment, light intensity was a limiting factor for photosynthesis. Write down the evidence used by the scientist to form this conclusion. **[1 mark]**

09.5 When the light is placed a distance of 0.10 m from the pondweed, carbon dioxide becomes the limiting factor. Sketch a second graph of light intensity against rate of photosynthesis to show this effect for lamp distances of 0.10–0.60 m. **[3 marks]**

10 Plants make their own food by the process of photosynthesis.

10.1 Write down the word equation for photosynthesis. **[2 marks]**

10.2 **Figure 8** shows the effect of temperature on the rate of photosynthesis.

Figure 8

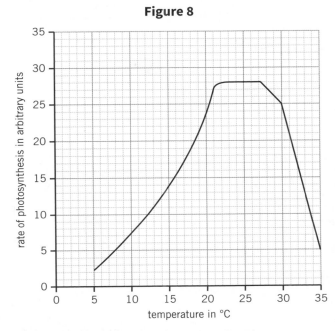

Using **Figure 8**, write down the range of optimum temperatures for photosynthesis. **[1 mark]**

10.3 Suggest why the rate of photosynthesis stays the same between these temperatures. **[2 marks]**

10.4 A farmer decides to use this information to set their greenhouse to the optimum conditions. Identify the best temperature to heat the farmer's greenhouse to. Give reasons for your answer. **[3 marks]**

11 Plants make their own food by respiration.

11.1 Complete the chemical reaction for photosynthesis. **[1 mark]**

$$6CO_2 + 6H_2O \rightarrow$$

11.2 Describe the role of chloroplasts in photosynthesis. **[3 marks]**

11.3 Large-scale commercial growers maximise the rate of photosynthesis in condition-controlled greenhouses. A farmer investigates how increasing the carbon dioxide concentration in a greenhouse affects their crop yield. The results are shown in **Table 3**. Carbon dioxide in the atmosphere is at 0.05 %.

Table 3

% carbon dioxide	Lettuce yield in kg/m²	Tomato yield in kg/m²
0.05	0.9	4.4
0.10	1.2	6.4
0.15	1.4	7.0
0.20	1.5	7.4

Explain how changing the carbon dioxide concentration changes the yield of lettuce and tomato crops. **[4 marks]**

11.4 To increase the carbon dioxide concentration by 0.05%, it costs the farmer £0.64 per m². The farmer sells their lettuce for £0.50/kg and their tomatoes for £1.60/kg. Evaluate the economic benefits of increasing carbon dioxide to 0.10%. **[6 marks]**

12 **Figure 9** shows the number of people who contracted measles in the UK between 1996 and 2008.

Figure 9

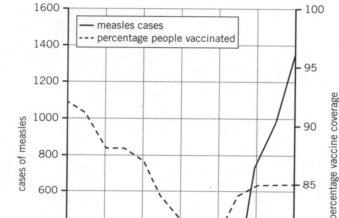

To protect people from contracting measles, children in the UK are offered the measles vaccination.

12.1 Explain how the measles vaccination prevents children from contracting measles. **[6 marks]**

12.2 Using **Figure 9**, evaluate the effectiveness of the vaccine. **[4 marks]**

12.3 A student concluded that herd immunity for measles is achieved with a vaccination rate of 85%. Justify the extent to which you agree, or disagree, with this conclusion. **[4 marks]**

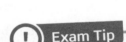

13 **Figure 10** shows some mitochondria, labelled in blue, from a liver cell.

13.1 Name the type of microscope which was used to produce this image. **[1 mark]**

13.2 The real length of the mitochondrion labelled 'X' is 10 μm. Using **Figure 10**, calculate the magnification used to produce this image. **[2 marks]**

$$\text{magnification} = \frac{\text{size of image}}{\text{size of real object}}$$

1 μm = 1000 mm

Figure 10

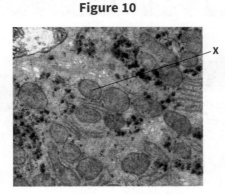

X

! **Exam Tip**

Don't worry if you've never seen a cell coloured like this before.

These are not the real colours, they just function to show the different parts of the cell.

14 **Figure 11** shows a diagram of the HIV virus. **Figure 12** shows a diagram of a *Salmonella* bacterium.

Figure 11

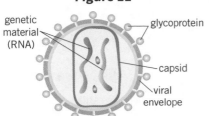

genetic material (RNA), glycoprotein, capsid, viral envelope

Figure 12

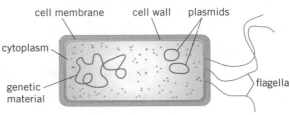

cell membrane, cell wall, plasmids, cytoplasm, genetic material, flagella

14.1 Give **one** structural feature that tells you **Figure 11** is a viral cell. **[1 mark]**

14.2 Give **one** structural feature that tells you **Figure 12** is a prokaryotic cell. **[1 mark]**

14.3 HIV is approximately 120 nm in diameter. *Salmonella* bacteria are approximately 5 μm in length. Calculate the order of magnitude of the size difference between a HIV virus and a *Salmonella* bacterium. **[1 mark]**

14.4 Many scientists do not classify viruses as being alive. Suggest why this conclusion has been formed. **[3 marks]**

B12 Respiration

Cellular respiration

Cellular **respiration** is an **exothermic** reaction that occurs continuously in the **mitochondria** of living cells to supply the cells with energy.

The energy released during respiration is needed for all living processes, including

- chemical reactions to build larger molecules, for example, making proteins from amino acids
- muscle contraction for movement
- keeping warm.

Respiration in cells can take place aerobically (using oxygen) or anaerobically (without oxygen).

Type of respiration	Oxygen required?	Relative amount of energy transferred
aerobic	✓	complete **oxidation** of glucose – large amount of energy is released
anaerobic	✗	incomplete oxidation of glucose – much less energy is released than in aerobic respiration

Aerobic respiration

glucose + oxygen → carbon dioxide + water Ⓛ

$$C_6H_{12}O_6 + 6O_2 \rightarrow 6CO_2 + 6H_2O$$

Anaerobic respiration in muscles

glucose → lactic acid

$$C_6H_{12}O_6 \rightarrow 2C_3H_6O_3$$

Fermentation

Anaerobic respiration in plant and yeast cells is represented by the equation: Ⓛ

glucose → ethanol + carbon dioxide

Anaerobic respiration in yeast cells is called **fermentation**.

The products of fermentation are important in the manufacturing of bread and alcoholic drinks.

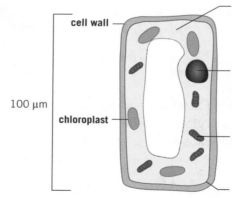

cell wall

cytoplasm
Where enzymes are made. Location of reactions in anaerobic respiration.

nucleus
Holds genetic code for enzymes involved in respiration.

100 µm

chloroplast

mitochondrion
Contains the enzymes for aerobic respiration.

cell membrane
Allows gases and water to pass freely into and out of the cell. Controls the passage of other molecules.

Typical plant cell

Typical animal cell

🐾 Revision tip

You need to learn the balanced symbol equations for the different types of respiration as well as the word equations.

🔑 Key terms

Make sure you can write a definition for these key terms.

exothermic   fermentation   lactic acid   metabolism

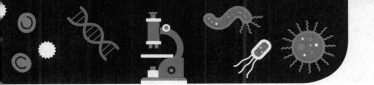

Response to exercise

During exercise the human body reacts to the increased demand for energy.

To supply the muscles with more oxygenated blood, heart rate, breathing rate, and breath volume all increase.

If insufficient oxygen is supplied, anaerobic respiration takes place instead, leading to the build-up of **lactic acid**.

During long periods of vigorous exercise, muscles become fatigued and stop contracting efficiently.

After exercise, the lactic acid accumulated during anaerobic respiration needs to be removed. **Oxygen debt** is the amount of oxygen needed to react with the lactic acid to remove it from cells.

Removal of lactic acid

lactic acid in the muscles

⬇

transported to the liver in the blood

⬇

lactic acid is converted back to glucose

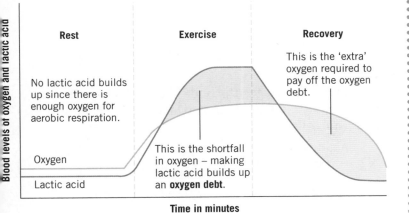

Metabolism

Metabolism is the sum of all the reactions in a cell or the body.

The energy released by respiration in cells is used for the continual enzyme-controlled processes of metabolism that produce new molecules.

Metabolic processes include the synthesis and breakdown of:

Carbohydrates

- synthesis of carbohydrates from sugars (starch, glycogen, and cellulose)
- breakdown of glucose in respiration to release energy

Proteins

- synthesis of amino acids from glucose and nitrate ions
- amino acids used to form proteins
- excess proteins broken down to form urea for excretion

Lipids

- synthesis of lipids from one molecule of glycerol and three molecules of fatty acid

 Revision tip

A question on respiration could be easily linked to one on enzyme action and break down of carbohydrates, proteins, or lipids.

Don't expect questions in the exam to cover only one topic, as they can link a few topics together within one question.

mitochondria   oxidation   oxygen debt   respiration

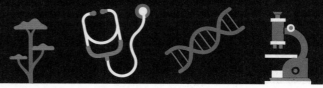
Learn the answers to the questions below then cover the answers column with a piece of paper and write as many as you can. Check and repeat.

B12 questions | Answers

	B12 questions	Answers
1	Define the term cellular respiration.	an exothermic reaction that occurs continuously in the mitochondria of living cells to release energy from glucose
2	What do organisms need energy for?	• chemical reactions to build larger molecules • muscle contraction for movement • keeping warm
3	What is the difference between aerobic and anaerobic respiration?	aerobic respiration uses oxygen, anaerobic respiration does not
4	Write the word equation for aerobic respiration.	glucose + oxygen → carbon dioxide + water
5	Write the word equation for anaerobic respiration in muscles.	glucose → lactic acid
6	Write the balanced symbol equation for aerobic respiration.	$C_6H_{12}O_6 + 6O_2 \rightarrow 6CO_2 + 6H_2O$
7	Why does aerobic respiration release more energy than anaerobic respiration?	oxidation of glucose is complete in aerobic respiration and incomplete in anaerobic respiration
8	What is anaerobic respiration in yeast cells called?	fermentation
9	Write the word equation for anaerobic respiration in plant and yeast cells.	glucose → ethanol + carbon dioxide
10	How does the body supply the muscles with more oxygenated blood during exercise?	heart rate, breathing rate, and breath volume increase
11	What substance builds up in the muscles during anaerobic respiration?	lactic acid
12	What happens to muscles during long periods of activity?	muscles become fatigued and stop contracting efficiently
13	What is oxygen debt?	amount of oxygen the body needs after exercise to react with the accumulated lactic acid and remove it from cells
14	How is lactic acid removed from the body?	lactic acid in muscles → blood transports to the liver → lactic acid converted back to glucose
15	What is metabolism?	sum of all the reactions in a cell or the body

Put paper here

Now go back and use the questions below to check your knowledge from previous chapters.

Previous questions	Answers
How are the lungs adapted for efficient gas exchange?	• alveoli – large surface area • moist membranes – increases rate of diffusion • one-cell-thick membranes – short diffusion pathway • good blood supply – maintains a steep concentration gradient
Name four types of pathogen.	bacteria, fungi, protists, viruses
What non-specific systems does the body use to prevent pathogens getting into it?	skin; cilia and mucus in the nose, trachea, and bronchi; stomach acid
Define health.	state of physical and mental well-being
Give the word equation for photosynthesis.	carbon dioxide + water $\rightarrow$ glucose + oxygen
How do plants use the glucose produced in photosynthesis?	• respiration • convert it into insoluble starch for storage • produce fat or oil for storage • produce cellulose to strengthen the cell wall • produce amino acids for protein synthesis
Give the balanced symbol equation for photosynthesis.	$6CO_2 + 6H_2O \rightarrow C_6H_{12}O_6 + 6O_2$

 ## Maths Skills

Practise your maths skills using the worked example and practice questions below.

Surface area-to-volume ratio	Worked Example	Practice
Knowledge of surface area-to-volume ratio is important in biology, for example, it explains the body size adaptations of organisms, and is important for the rate at which transportation process such as respiration occur. To calculate it, you first need to calculate the surface area and volume of the object. For the surface area of a cube, find the area of one face and multiply by six. To find the volume of a cube, use: length × width × height To calculate surface area to volume ratio: surface area-to-volume ratio $= \dfrac{\text{surface area}}{\text{volume}}$	What is the surface area-to-volume ratio of the cube below? 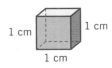 1 cm 1 cm 1 cm To calculate surface area (cm²): Area of one side of the cube = 1 × 1 = 1 cm² The cube has six sides, so: surface area = 1 × 6 = 6 cm² To calculate volume (cm³): 1 × 1 × 1 = 1 cm³ Surface area-to-volume ratio: $\dfrac{6}{1}$ = 6:1 ratio	Work out the surface area (cm²), volume (cm³), and surface area-to-volume ratio for cubes with the following dimensions. **1** 2 cm × 2 cm × 2 cm **2** 5 cm × 5 cm × 5 cm **3** 12 cm × 12 cm × 12 cm

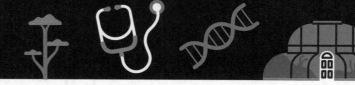

Practice

Exam-style questions

01 Animals transfer energy into a form that cells can use by the process of respiration.

01.1 Complete the following balanced chemical equation to summarise the process of aerobic respiration in animal cells. **[2 marks]**

$$6\,O_2 + \underline{\hspace{3cm}} \rightarrow \underline{\hspace{3cm}} + 6\,CO_2 + energy$$

> **!** **Exam Tip**
>
> Instead of remembering the chemical formula and then trying to balance this equation, it's easier to remember the numbers needed to balance it.

01.2 Explain why respiration is described as an exothermic reaction. **[3 marks]**

_____

_____

_____

_____

01.3 Explain why fat cells do not have as many mitochondria as muscle cells. **[3 marks]**

> **!** **Exam Tip**
>
> Think about the differences in function between muscle and fat cells.

_____

_____

_____

_____

01.4 Plants do not move or maintain a certain body temperature.
Describe **two** ways plants use energy from respiration. **[2 marks]**

1 _____

_____

2 _____

_____

02 A group of students investigated the effect of temperature on the rate of aerobic respiration in earthworms.

They placed the equipment shown in **Figure 1** into a water bath.

Figure 1

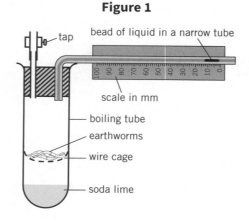

- tap
- bead of liquid in a narrow tube
- scale in mm
- boiling tube
- earthworms
- wire cage
- soda lime

02.1 Identify the function of soda lime. **[1 mark]**
Tick **one** box.

to absorb carbon dioxide ✓

to absorb oxygen ☐

to provide the earthworms with water ☐

to provide the earthworms with nutrients ☐

02.2 Explain how this equipment can be used to measure the rate of respiration at different temperatures. **[6 marks]**

_____

_____

_____

_____

_____

_____

_____

_____

> **! Exam Tip**
>
> You'll need a safe way to heat the boiling tube that can be controlled and measured.

02.3 Write down **one** ethical consideration needed in this experiment. **[1 mark]**

_____

_____

02.4 The students' results at 10 °C and 20 °C are shown in **Figure 2**.

Figure 2

Compare the rate of respiration of earthworms at 10 °C and 20 °C. **[5 marks]**

_____

_____

_____

_____

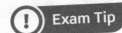

Exam Tip

This question is worth five marks, so you need to include data from both lines on the graph, and talk about the differences between the lines.

02.5 Draw a line on **Figure 2** to show what you predict the rate of respiration would be for earthworms at 25 °C.

Give reasons for your answer. **[3 marks]**

_____

_____

_____

_____

03 Yeast respires anaerobically. This reaction is called fermentation and it is used in the manufacture of some foods.

03.1 Write down the word equation for anaerobic respiration in yeast cells. **[2 marks]**

Exam Tip

Both of the products of this reaction are used in the food industry.

03.2 To make food products efficiently, producers need to know the optimum conditions for yeast to respire. A food scientist set up the apparatus in **Figure 3** to study how temperature affects yeast respiration.

Figure 3

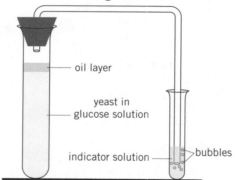

The indicator solution responds as follows:

blue   green   yellow

very low ————————————————→ high

carbon dioxide concentration

03.2 Suggest which colour the indicator will turn when yeast is respiring at its maximum rate. **[1 mark]**

03.3 Explain the purpose of the layer of oil. **[1 mark]**

Exam Tip

The important thing about the oil is that it is sitting on top of the mixture.

03.4 Explain how you would use this equipment to study the effect of temperature on the rate of respiration. **[4 marks]**

03.5 Suggest how you can adapt this investigation to obtain quantitative data on the rate of yeast respiration. **[3 marks]**

04 **Figure 4** shows the concentration of lactic acid in a person's blood. The concentration of lactic acid was measured before, during, and after ten minutes of vigorous exercise.

Figure 4

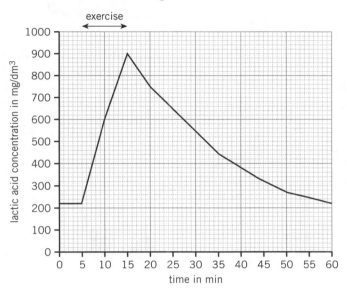

04.1 Identify the level of lactic acid in the person's blood before exercise. **[1 mark]**

04.2 Explain why lactic acid was produced between 5 and 15 minutes. **[2 marks]**

04.3 Calculate the rate at which lactic acid was produced during the period of exercise. **[2 marks]**

04.4 Describe and explain the trend shown by the graph between 15 and 60 minutes. **[6 marks]**

05 Respiration takes place in all living plant and animal cells.

05.1 Describe the purpose of respiration. **[2 marks]**

05.2 Organisms can respire both aerobically and anaerobically. **Table 1** summarises the similarities and differences between these processes. Complete the table. **[4 marks]**

> **! Exam Tip**
>
> Don't forget to give the units!

> **! Exam Tip**
>
> The clue to this answer is in the question.

> **! Exam Tip**
>
> Draw a large triangle on the graph to help you calculate this. Remember to write down your working out.

Table 1

Type of respiration	aerobic	anaerobic	
Organism it occurs in	plants and animals	plants	animals
Oxygen required?	yes	no	
Glucose required?	yes		yes
Carbon dioxide produced?	yes	yes	
Other products produced	water		lactic acid

05.3 Describe **two** ways yeast is used to produce food products using anaerobic respiration. **[4 marks]**

06 **Figure 5** shows the main components in an animal cell as seen under a light microscope.

06.1 Identify the part of the cell where respiration occurs. **[1 mark]**

06.2 What are the products of respiration? Choose **one** answer. **[1 mark]**

glucose + oxygen

glucose + carbon dioxide

carbon dioxide + water

water + oxygen

Figure 5

! Exam Tip

The equations for respiration is the opposite to photosynthesis, so if you have trouble write down the one you remember and reverse it.

06.3 When a person is exercising vigorously, body cells can switch to anaerobic respiration. Give **two** reasons why animal cells normally respire aerobically. **[2 marks]**

07 As part of a fitness test, an athlete ran as fast as possible for 15 minutes on a treadmill. The glucose and lactic acid concentrations of the athlete's blood were measured. **Figure 6** shows the results.

Figure 6

concentration in mmol/dm³

20

15

10

5

0

start end
glucose

start end
lactic acid

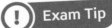

07.1 Using **Figure 6**, explain how the athlete respired over the 15 minutes of their fitness test. **[4 marks]**

! Exam Tip

You'll need to use data from the graph to get full marks on this question.

07.2 Calculate the percentage change in the athlete's lactic acid concentration between the start and end of the exercise. **[2 marks]**

! Exam Tip

Show all of your working out.

07.3 Explain why the athlete's heart and breathing rates increased during the fitness test. **[6 marks]**

08 All living organisms metabolise chemical compounds.

08.1 Define the term metabolism. **[1 mark]**

08.2 Explain how respiration maintains the rate of metabolism in an organism. **[3 marks]**

08.3 Two examples of metabolic reactions are the formation of lipids and the conversion of glucose into starch (in plants) or glycogen (in animals). Name the molecules required to form a lipid. **[2 marks]**

! Exam Tip

Potatoes are an example of a starch.

08.4 Plants convert glucose into starch, and animals convert glucose into glycogen. Explain why this is necessary. **[2 marks]**

09 Endurance athletes wish to avoid lactic acid build-up in their muscle cells.

09.1 Lactic acid build-up is a concern for marathon runners, but is not an important consideration for sprinters. Explain why. **[4 marks]**

09.2 To increase the blood's oxygen-carrying capability, many marathon runners train at high altitudes. This encourages the body to produce more red blood cells. Explain why high-altitude training can improve the performance of an athlete. **[4 marks]**

09.3 Blood doping is an illegal practice that mimics altitude training. An athlete provides up to two litres of blood several weeks before a competition, which is then stored. The blood is then re-infused into the athlete one week before the competition. Explain why blood doping would produce a performance enhancement for an athlete similar to the effect produced through high-altitude training. **[3 marks]**

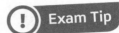

10 All living organisms respire.

10.1 Complete the chemical equation for aerobic respiration. **[1 mark]**

$C_6H_{12}O_6 +$ _____ $\rightarrow 6CO_2 +$ _____

10.2 Explain **two** reasons why animals usually respire aerobically. **[4 marks]**

10.3 Cheetahs are able to sprint very fast. After sprinting a cheetah will puff and pant for several minutes. With reference to respiration, explain why this happens. **[4 marks]**

11 Variegated leaves are leaves with areas that do not contain chlorophyll. To show that chlorophyll is needed for a plant to photosynthesise, a scientist tested a leaf for the presence of starch.

11.1 Explain why the scientist first boiled the leaf in ethanol and then washed the leaf using water. **[3 marks]**

11.2 Describe and explain **one** safety procedure the scientist should follow when performing this experiment. **[2 marks]**

11.3 Predict and explain the results the scientist will observe when iodine is added to different areas of the leaf. **[4 marks]**

11.4 Explain how a plant produces starch. **[6 marks]**

12   **Figure 7** shows some of the main structures in the respiratory system.

Figure 7

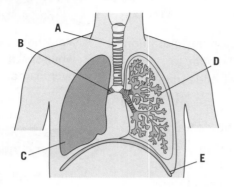

12.1   Identify the parts of the respiratory system labelled **A–E**. **[5 marks]**

12.2   Explain how the alveoli are adapted for gas exchange.   **[3 marks]**

12.3   Emphysema is an example of a lung disease. The walls of the alveoli break down, forming larger air spaces than normal. A doctor measured the volume of air in the lungs of two people over a period of eight seconds. The measurements were taken at rest.

Table 2

Time in seconds	Volume of air in lungs in dm³	
	Person 1	Person 2
0	6.8	7.2
1	4.0	6.2
2	2.8	5.8
3	2.3	5.3
4	2.0	5.0
5	1.8	4.8
6	1.5	4.5
7	1.5	4.2
8	1.5	3.9

Use evidence from **Table 2** to explain whether the doctor measured the two people inhaling or exhaling.   **[2 marks]**

12.4   Compare the mean rate of change of lung volume in the two people over the eight seconds of the test.   **[5 marks]**

12.5   One person in the investigation had healthy lungs. The other person had emphysema. Identify which person has emphysema, giving reasons for your answer.   **[2 marks]**

> ! **Exam Tip**
>
> Look at the difference in the start and end values.

> ! **Exam Tip**
>
> Remember to give units for your answer.

13   Organisms are made up of a number of organ systems. One
       example is the digestive system.

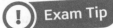

13.1 Give another example of an organ system and describe
       its function.   **[2 marks]**

13.2 The main organs in the digestive system are labelled in **Figure 8**.
       Identify the label that shows the liver.   **[1 mark]**

Figure 8

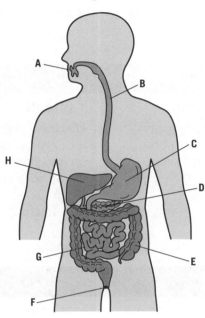

13.3 Describe the function of organ **E**.   **[1 mark]**

13.4 Explain **two** reasons why organ **C** contains hydrochloric acid.
                                                            [2 marks]

14   Many people donate blood. The donated blood can be used to
       treat a number of medical conditions or to replace blood lost in an
       accident.

14.1 Explain why blood is described as a tissue.   **[1 mark]**

14.2 The largest single component of blood is blood plasma. Blood
       cells and platelets are transported around the body by the plasma.
       Name **two** other substances that are transported around the body
       in the plasma.   **[2 marks]**

14.3 Explain the role of white blood cells in the body's defence
       against pathogens.   **[6 marks]**

For answers and more practice questions visit
www.oxfordrevise.com/scienceanswers   Even more practice and interactive
                                             revision quizzes are available on (kerboodle)   **B12 Practice**   141

B13 Nervous system and homeostasis A

The nervous system

Function

The nervous system enables humans to react to their surroundings and to coordinate their behaviour – this includes both voluntary and **involuntary** actions.

Structure

The nervous system is made up of the **central nervous system** (CNS) and a network of nerves.

The CNS comprises the **brain** and **spinal cord**.

Nervous system responses

Stimulus	**Receptor**	**Coordinator**	**Effector**	**Response**
a change in the environment (**stimulus**) is detected by **receptors**	information from receptors passes along cells (**neurones**) to the CNS as electrical impulses	the CNS coordinates the body's response to the stimulus	effectors bring about a response, such as glands secreting hormones or muscles contracting	the body responds to the stimulus

Reflex arcs

Reflex actions of the nervous system are automatic and rapid – they do not involve the conscious part of the brain.

Reflex actions are important for survival because they help prevent damage to the body.

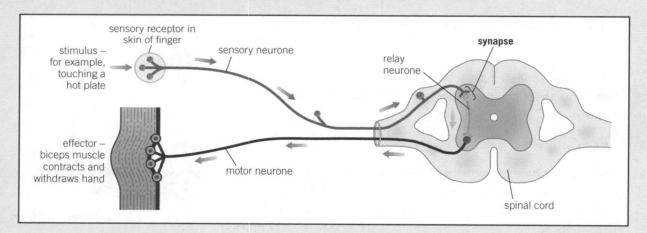

 Key terms

Make sure you can write a definition for these key terms.

| brain | central nervous system | effectors | involuntary | neurones |
| receptors | reflex action | spinal cord | stimulus | synapse |

Paper 2 starts here

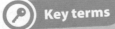

Reflex arc structures

Neurones

carry electrical impulses around the body – relay neurones connect sensory neurones to motor neurones

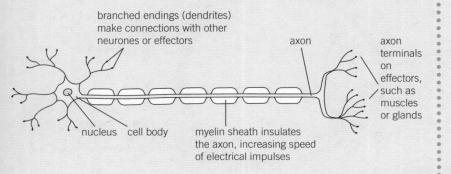

branched endings (dendrites) make connections with other neurones or effectors

axon

axon terminals on effectors, such as muscles or glands

nucleus   cell body

myelin sheath insulates the axon, increasing speed of electrical impulses

Synapses

gaps between neurones, which allow electrical impulses in the nervous system to cross between neurones

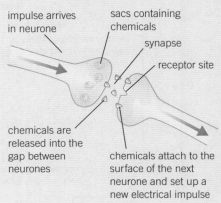

impulse arrives in neurone

sacs containing chemicals

synapse

receptor site

chemicals are released into the gap between neurones

chemicals attach to the surface of the next neurone and set up a new electrical impulse

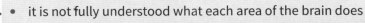

The brain

The brain controls complex behaviour.

It is made of billions of interconnected neurones, with different regions that carry out different functions.

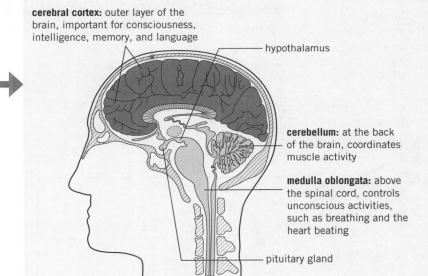

cerebral cortex: outer layer of the brain, important for consciousness, intelligence, memory, and language

hypothalamus

cerebellum: at the back of the brain, coordinates muscle activity

medulla oblongata: above the spinal cord, controls unconscious activities, such as breathing and the heart beating

pituitary gland

Research on the brain

Neuroscientists have mapped the regions of the brain to particular functions by studying patients with brain damage, using MRI scanning techniques, and electrically stimulating parts of the brain.

The brain is very complex and delicate, making investigating and treating brain disorders difficult.

Brain damage and diseases can involve many different neurones, chemicals, and areas of the brain. Treatment is difficult because

- it is not fully understood what each area of the brain does
- drugs do not always reach the brain through its membranes
- surgery can easily cause unintended damage.

Knowledge

B13 Nervous system and homeostasis B

Structure of the eye

The eye is a **sense organ** containing **receptors** sensitive to light intensity and colour.

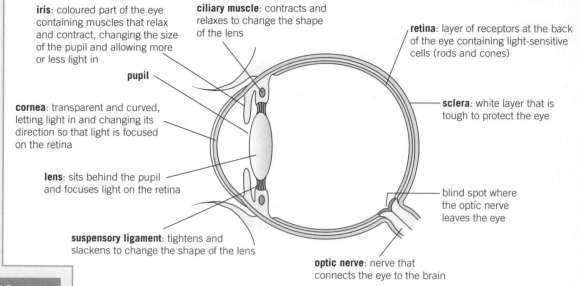

iris: coloured part of the eye containing muscles that relax and contract, changing the size of the pupil and allowing more or less light in

pupil

cornea: transparent and curved, letting light in and changing its direction so that light is focused on the retina

lens: sits behind the pupil and focuses light on the retina

suspensory ligament: tightens and slackens to change the shape of the lens

ciliary muscle: contracts and relaxes to change the shape of the lens

retina: layer of receptors at the back of the eye containing light-sensitive cells (rods and cones)

sclera: white layer that is tough to protect the eye

blind spot where the optic nerve leaves the eye

optic nerve: nerve that connects the eye to the brain

Accommodation

Accommodation is the process of changing the shape of the lens to focus on near or distant objects.

To focus on a *near* object
- ciliary muscles *contract*
- suspensory ligaments are slack
- so lens is thicker and more curved, and refracts light rays more strongly.

• •

To focus on a *distant* object
- ciliary muscles relax
- suspensory ligaments are pulled tight
- so lens is thinner and flatter, and only refracts light rays slightly.

Common defects of the eyes

Myopia

Short-sightedness, when distant objects look blurred because rays of light focus in front of the retina.

This is corrected using **concave** spectacle lenses.

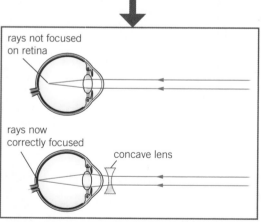

rays not focused on retina

rays now correctly focused

concave lens

Hyperopia

Long-sightedness, when near objects look blurred because rays of light focus behind the retina.

This is corrected using **convex** spectacle lenses.

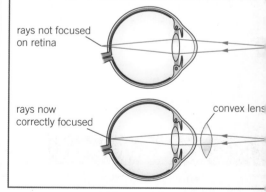

rays not focused on retina

rays now correctly focused

convex lens

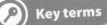 **Key terms**

Make sure you can write a definition for these key terms.

accommodation   concave   convex   coordination centre   effector   homeostasis   hyperopia   myopia

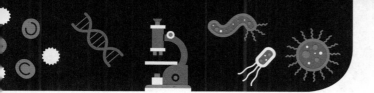
Homeostasis

Homeostasis is the regulation of internal conditions (of a cell or whole organism) in response to internal and external changes, to constantly maintain optimum conditions for functioning.

This maintains optimum conditions for all cell functions and enzyme action.

In the human body, this includes control of

- blood glucose concentration
- body temperature
- water levels.

The automatic control systems of homeostasis may involve nervous responses or chemical responses.

All control systems involve

- receptor cells, which detect **stimuli** (changes in the environment)
- **coordination centres** (such as the brain, spinal cord, and pancreas), which receive and process information from receptors
- **effectors** (muscles or glands), which produce responses to restore optimum conditions.

Control of body temperature

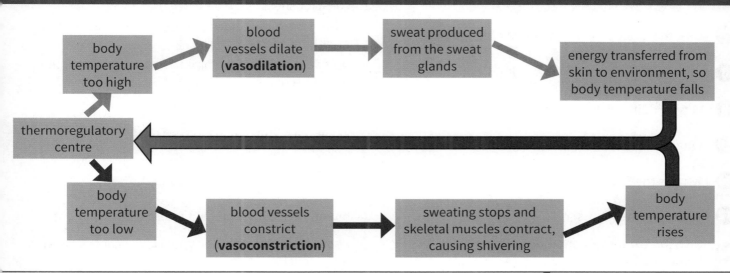

Treatment of eye defects

- spectacle lenses to refract light rays to focus on the retina
- hard and soft contact lenses – like traditional glasses, but on the surface of the eye
- laser eye surgery – to change the shape of the cornea
- replacement lenses – adding another lens inside the eye to correct defects permanently.

Body temperature is monitored and controlled by the **thermoregulatory centre** in the brain. The centre contains receptors sensitive to the blood temperature.

The skin also contains temperature receptors and sends nervous impulses to the thermoregulatory centre.

receptor   sense organ   thermoregulatory centre   vasoconstriction   vasodilation

Retrieval

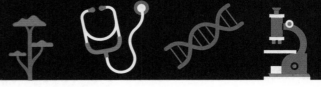

Learn the answers to the questions below then cover the answers column with a piece of paper and write as many as you can. Check and repeat.

B13 questions		Answers
1	What is the function of the nervous system?	it enables organisms to react to their surroundings and coordinates behaviour
2	What are the two parts of the central nervous system?	brain and spinal cord
3	Why are reflex actions described as rapid and automatic?	they do not involve the conscious part of the brain
4	Why are reflex actions important?	for survival and to prevent damage to the body
5	Give the pathway of a nervous response.	stimulus → receptor → coordinator → effector → response
6	Give the function of the cerebral cortex.	outer layer of the brain playing an important role in consciousness
7	Give the function of the medulla oblongata.	part of the brain above the spinal cord that controls breathing and heart rate
8	Give the function of the cerebellum.	part at the back of the brain involved in coordinating muscle activity
9	Why is it difficult to treat brain disorders?	brain is very complex and delicate
10	What is a synapse?	gap between two neurones, allowing impulses to cross
11	What is the function of neurones?	carry electrical impulses around the body
12	What is homeostasis?	maintenance of a constant internal environment
13	Give three internal conditions controlled in homeostasis.	body temperature, blood glucose concentration, and water levels
14	Give three things all control systems include.	receptors, coordination centres, and effectors
15	What is accommodation?	process of changing the shape of the lens to focus on near/distant objects
16	Give two common defects of the eyes.	myopia (short-sightedness) and hyperopia (long-sightedness)
17	How can eye defects be treated?	spectacle lenses, contact lenses, laser surgery, and replacement lenses in the eye
18	Where is body temperature monitored and controlled?	thermoregulatory centre in the brain
19	What happens if body temperature is too high?	blood vessels dilate (vasodilation) and sweat is produced
20	What happens if body temperature is too low?	blood vessels constrict (vasoconstriction), sweating stops, and shivering takes place

Put paper here

Now go back and use the questions below to check your knowledge from previous chapters.

B13

Previous questions | Answers

Previous questions		Answers
Define the term cellular respiration.	Put paper here	an exothermic reaction that occurs continuously in the mitochondria of living cells to release energy from glucose
Give the equation for the inverse square law for light intensity.		light intensity $\propto \dfrac{1}{\text{distance}^2}$
How is lactic acid removed from the body?	Put paper here	lactic acid in muscles → blood transports to the liver → lactic acid converted back to glucose
How does the body supply the muscles with more oxygenated blood during exercise?		heart rate, breathing rate, and breath volume increase
Describe how light intensity affects the rate of photosynthesis.	Put paper here	increasing light intensity increases the rate of photosynthesis until another factor becomes limiting
Why should an inoculating loop be passed through a blue Bunsen flame before and after use?		sterilise it/kill any bacteria
What is coronary heart disease?	Put paper here	layer of fatty material that builds up inside the coronary arteries, narrowing them – results in a lack of oxygen for the heart

Required Practical Skills

Practise answering questions on the required practicals using the example below.
You need to be able to apply your skills and knowledge to other practicals too.

Reaction times

You need to be able to describe how to plan an experiment and choose suitable variables to change, to look at how different variables affect reaction times.

You should be able to:

- write hypotheses predicting the effects of changing single variables

- be able to identify independent and dependent variables

- evaluate results in terms of accuracy and precision

- understand how different factors affect human reaction times.

Worked example

Write a method to test the effect of caffeine consumption on human reaction time.

Answer: With two people working in pairs, have the first person hold a ruler vertically with zero at the bottom. The second person should steady their arm on the edge of a bench underneath the ruler. The first person should drop the ruler, and the second person should catch the ruler between thumb and forefinger without moving their arm. Record the number on the ruler above the catcher's thumb. Repeat this at least three more times. The second person should then consume a caffeinated drink, and the entire experiment repeated.

Practice

1. Give two things that must be controlled for the investigation of caffeine on reaction time to be a fair test.

2. Explain why it would not be appropriate for the same person to drop and catch the ruler in an experiment on reaction times.

3. Why should the test of reaction time be repeated multiple times before and after consumption of the caffeinated drink?

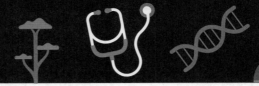

Exam-style questions

01 A student carried out an investigation to determine whether a person's reaction time was quicker with their dominant hand.

The student used the following steps:

1 The student investigator (Student **A**) held a ruler just above a second student's hand (Student **B**).

2 Student **A** let go of the ruler. Student **B** caught it as soon as possible.

3 The experiment was then repeated with Student **B**'s opposite hand.

> **! Exam Tip**
>
> Your dominant hand is the one you write with!

01.1 Identify the dependent variable in this investigation. **[1 mark]**

Reaction time.

01.2 Identify **one** variable that should be controlled in this investigation. **[1 mark]**

Students coffine or alcohol intake.

Student **A** chose ten right-handed students to test. The results are shown in **Table 1**.

Table 1

Student		1	2	3	4	5	6	7	8	9	10	Mean
Reaction time in s	Left hand	0.25	0.23	0.39	0.26	0.27	0.22	0.25	0.27	0.25	0.25	0·225
	Right hand	0.28	0.24	0.25	0.27	0.26	0.22	0.26	0.24	0.23	0.25	0.25

01.3 Identify the anomalous result from the experiment. **[1 mark]**

0.39

01.4 Complete **Table 1** by calculating the mean reaction time for the left hand results. **[1 mark]**

Mean = ___0.225___ s

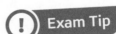

> **! Exam Tip**
>
> Whenever you're asked to calculate a mean, look out for any anomalous results and exclude them.

01.5 Student **A** reached the following conclusion: 'Right-handed people's reaction times are more rapid when using their dominant (right) hand.'

Explain the extent to which you agree or disagree with this conclusion. **[3 marks]**

The statement is true for roughly 50% of the students so I agree to an extent of 50% with this conclusion.

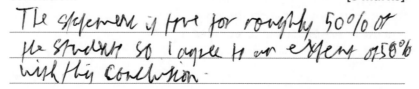

01.6 Which type of reaction time did the experiment measure?
Tick **one** box. **[1 mark]**

reflex reaction time ☐

voluntary reaction time ☐

01.7 Give a reason for your answer to **01.6**. **[1 mark]**

_____

_____

01.8 The fastest measured reaction time of a human is approximately
0.1 s. Explain why a person's reaction time cannot be more rapid
than this. **[3 marks]**

_____

_____

_____

_____

02 To keep an organism healthy both the nervous system and the
hormonal system need to work together.

02.1 Define the term homeostasis. **[1 mark]**

_____

_____

02.2 One purpose of homeostasis is to provide optimal conditions for
enzymes to work in.

Name **two** factors that can affect the rate of enzyme
action. **[2 marks]**

1 _____

2 _____

02.3 Homeostasis involves a number of automatic control systems.

Explain why they are described as automatic systems. **[1 mark]**

_____

_____

02.4 Describe the main components of a control system. **[6 marks]**

! **Exam Tip**

Go through each part one at a time.

_____

_____

_____

_____

_____

_____

_____

_____

_____

_____

03 **Figure 1** shows a drawing of the brain.

Figure 1

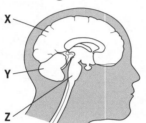

03.1 Identify parts **Y** and **Z** in **Figure 1**. **[2 marks]**

03.2 A climber hurts their head in a climbing accident and is taken to hospital. The climber cannot remember why they have been taken to hospital. Suggest which part of the brain the climber may have damaged. Identify the letter (**X**, **Y**, or **Z**) pointing to this part in **Figure 1**. **[2 marks]**

03.3 Name the imaging technique used to find out if the climber's brain has been damaged. **[1 mark]**

03.4 Explain why damage to the brain may be very difficult to treat. **[4 marks]**

04 The nervous system allows humans to respond to their surroundings. **Figure 2** shows two different nerve pathways joining a big toe to the central nervous system. A touch to the big toe causes an impulse to travel through the nerve pathway.

Figure 2

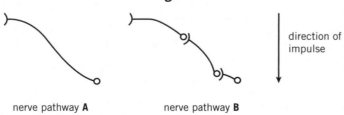

direction of impulse

nerve pathway **A**   nerve pathway **B**

04.1 Identify the type of neurone involved in this pathway. Choose **one** answer. **[1 mark]**

sensory neurone   motor neurone   relay neurone

04.2 Nerve pathway **A** is 90 cm long. A nerve impulse travels along this pathway at 76 m/s. Calculate how long it takes for the nerve impulse to travel the length of the pathway.

Give your answer to two significant figures. **[3 marks]**

04.3 Nerve pathways **A** and **B** are the same total length. The nerve impulse takes longer to travel along pathway **B** than pathway **A**. Use your knowledge and information in **Figure 2** to explain why. **[4 marks]**

05 **Figure 3** shows the cross-section of an eye.

Figure 3

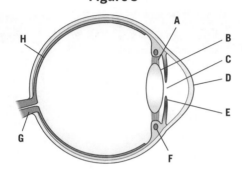

05.1 Using labels **A–H** from **Figure 3**, identify
- the suspensory ligaments
- the part of the eye that contains light-sensitive cells
- two parts of the eye that refract light. **[3 marks]**

05.2 A person moves from a dark room to a brightly lit room. Explain the changes that will occur in the eye as a result of this change. **[3 marks]**

05.3 This person looks out of the window at a tree in the distance. The person then looks down and picks up a book to read.

Explain the changes that occur in the eye to ensure this person is able to focus clearly on the distant tree, and then on the nearby book. **[4 marks]**

06 A student carried out an investigation into the reaction times of one person. The student measured the person's reaction time by dropping a ruler and noting the drop distance.

The student dropped the ruler five times and calculated the mean drop distance to be 115 mm.

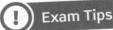

Exam Tips

The equation you need for this question is more often found in maths or physics, but there's no reason a question like this can't come up in biology.

Watch out for the non-standard units here.

Exam Tip

Look at the differences in structure.

Exam Tip

Labelling an eye is a question that frequently comes up in exams!

06.1 Use the following equation to calculate the person's reaction time. Give your answer to two significant figures. **[3 marks]**

$$\text{reaction time (s)} = \sqrt{\frac{\text{mean drop distance (cm)}}{490}}$$

Exam Tip

Don't worry if you've never seen this equation before. There will be some things in the exam that you've never seen – it's all about your skill at applying your biology knowledge to new situations!

06.2 The student wanted to determine if caffeine had an impact on reaction time.

Describe an approach the student could take to investigate this. **[4 marks]**

06.3 Predict and explain what you would expect to happen to a person's reaction time after they consume a caffeinated drink. **[3 marks]**

Exam Tip

The phrasing is the key to this question: *what* do you think will happen, and *why* do you think this?

06.4 The student extended the investigation to whether or not a person's dominant hand affects their reaction time. They collected the data shown in **Table 2**.

Table 2

Person tested	Repeat 1		Repeat 2	
	Left hand	Right hand	Left hand	Right hand
A (right-handed)	0.28	0.25	0.27	0.24
B (left-handed)	0.21	0.25	0.23	0.25

Exam Tip

'Evaluate' means you need to give an opinion and then justify that opinion.

The student concluded that person **A**'s reaction time is shorter when using their dominant hand to respond to a situation'.

Discuss and evaluate the validity of the student's conclusion.

[4 marks]

07 A student shouted loudly behind their friend. The friend jumped in reaction to the noise.

Explain in detail how the friend responded to the noise through the actions of their nervous system. **[6 marks]**

Exam Tip

You need to include the path the signal took and what happened at each stage.

08 As a car driver approached a set of traffic lights, the lights turned red. This caused the driver to press the brake pedal with their foot, slowing the car down.

08.1 In this response, what is the changing traffic light? Choose **one** answer. **[1 mark]**

the coordinator   the effector   the receptor   the stimulus

08.2 In this response, what is the coordination centre? Choose **one** answer. **[1 mark]**

the eye   the brain   a synapse   the spinal cord

08.3 Whilst waiting at the traffic lights, an insect flies close to the driver's eye. The driver's eye closes in response.

Explain how the driver's response to the insect is different to the response to the changing traffic lights. **[3 marks]**

09 If a person needs to have a dental procedure, such as a filling, a dentist will often inject the gum with an anaesthetic so the person does not feel any pain. Procaine is an example of an anaesthetic drug used by a dentist. **Figure 4** shows what happens at one of the synapses in your gum.

Figure 4

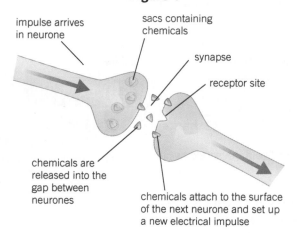

impulse arrives in neurone

sacs containing chemicals

synapse

receptor site

chemicals are released into the gap between neurones

chemicals attach to the surface of the next neurone and set up a new electrical impulse

09.1 Name the type of neurone that transmits the electrical impulse from the pain receptor. **[1 mark]**

09.2 Procaine is a competitive inhibitor. It is very similar to the chemical that is released between neurones.

Using your own knowledge and information from **Figure 4**, suggest how the drug procaine may work. **[6 marks]**

> (!) **Exam Tip**
>
> Lots of the information you need to answer this is in **Figure 4**.

10 This question is about defects in the eye.

10.1 Define the term accommodation in reference to the human eye. **[1 mark]**

> (!) **Exam Tip**
>
> You can't draw ray diagrams without a ruler!

10.2 Sketch a ray diagram to show why a person suffering from myopia is unable to see distant objects clearly. **[3 marks]**

10.3 Using your diagram from **10.2**, show how wearing glasses can be used to enable a myopic person to see a distant object clearly. **[3 marks]**

10.4 As people get older they often find it difficult to see nearby objects and need reading glasses.

Suggest and explain why this happens. **[4 marks]**

11 The nervous system controls the body's response to changes in its external environment.

11.1 Complete the following flow chart to name the main steps in a nervous response. **[1 mark]**

Stimulus → _____ → sensory neurone → CNS → motor neurone → _____

11.2 Name the part of the nervous system that the vertebral column protects. **[1 mark]**

11.3 Describe **two** differences between a motor neurone and a sensory neurone. **[2 marks]**

⚠ **Exam Tip**

'Describe' is what the neurones look like, not why they look this way.

11.4 Many sports require good reactions. In relation to the nervous system, explain what reaction time depends on. **[2 marks]**

12 A human's body temperature remains almost constant despite changes in external temperatures. When you walk outside on a cold day, your body is able to maintain an internal temperature of 37 °C.

⚠ **Exam Tip**

Think about the molecules needed to keep the body's chemical reactions happening.

12.1 Suggest what would happen if your body temperature dropped by 2 °C. **[2 marks]**

12.2 Explain how the body would detect and respond to a decrease in body temperature. **[6 marks]**

12.3 An athlete can run for two hours in temperatures of 35 °C if the air is dry without overheating. However, in humid conditions, if the temperature rises above 35 °C then their body will overheat. Suggest an explanation for the athlete overheating in humid conditions. **[3 marks]**

13 A group of people became ill after eating out in a restaurant. Health and safety inspectors carried out an investigation. The suspected source was food poisoning caused by bacteria on the rice. **Table 3** gives details of how four different restaurants stored their rice.

Table 3

Name of restaurant	Storage method	Time in storage in h	Storage temperature in °C
Tom's Diner	rice in an open container	8	30
George's Hot Dinners	rice left on a hot plate	1	50
Amira's Kitchen	rice frozen in a freezer compartment	24	−5
Betty's Home Cooking	rice in a sealed container	5	20

13.1 Suggest which restaurant the diners are most likely to have eaten in. **[1 mark]**

13.2 Explain your choice. **[2 marks]**

13.3 Name **one** way to treat food poisoning. **[1 mark]**

⚠ **Exam Tip**

Look at both the storage temperature and the time in storage.

13.4 Explain why uncooked rice can be stored for many months in a packet or jar without spoiling. **[1 mark]**

13.5 Some diseases are contagious. Suggest and explain **two** ways you could prevent a contagious disease being spread between individuals living in the same house. **[4 marks]**

14 Leaves are specially adapted for photosynthesis.

14.1 Write the balanced chemical equation for photosynthesis. **[2 marks]**

14.2 Explain **two** ways in which leaves are adapted to maximise the absorption of sunlight. **[4 marks]**

14.3 Stomata play an important role in the movement of gases involved in photosynthesis.

Compare and explain the net movement of carbon dioxide through the stomata at midday and midnight. **[6 marks]**

14.4 Stomata are usually found on a leaf's lower surface. This is not the case for many aquatic plants, such as water lilies, whose leaves float on the surface of the water.

Explain why the stomata of many aquatic plants are found on the leaf upper surface. **[4 marks]**

> **! Exam Tip**
>
> At midday and midnight different chemical reactions are happening. How do these differ in relation to carbon dioxide?

B14 Hormonal coordination A

Human endocrine system

The **endocrine system** is composed of glands that secrete chemicals called **hormones** into the bloodstream.

The blood carries hormones to a target organ, where an effect is produced.

Compared to the nervous system, the effects caused by the endocrine system are slower but act for longer.

The **pituitary gland**, located in the brain, is known as a 'master gland', meaning it secretes several hormones into the blood.

These hormones then act on other glands to stimulate the release of other hormones, and bring about effects.

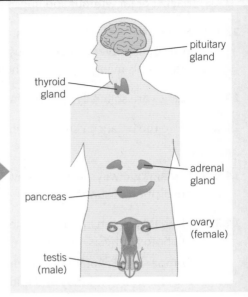

Control of blood glucose levels

Blood glucose (sugar) concentration is monitored and controlled by the **pancreas**.

This is an example of **negative feedback** control, as the pancreas switches production between the hormones **insulin** and **glucagon** to control blood glucose levels.

Diabetes

Diabetes is a non-communicable disease in which the body either cannot produce or respond to insulin, leading to uncontrolled blood glucose concentrations.

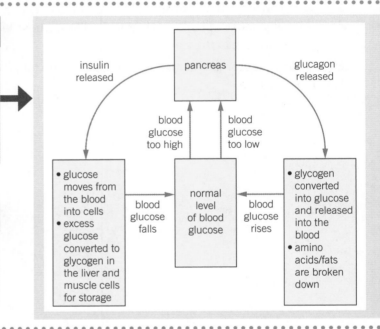

Type 1 diabetes	Type 2 diabetes
early onset	usually later onset, obesity is a risk factor
pancreas stops producing sufficient insulin	body doesn't respond to the insulin produced
commonly treated through insulin injections, also diet control and exercise	commonly treated through a carbohydrate-controlled diet and exercise

Key terms

Make sure you can write a definition for these key terms.

ADH   adrenal gland   adrenaline   diabetes   dialysis   endocrine system

glucagon   hormone   insulin   kidney tubule   metabolic rate   negative feedback

pancreas   pituitary gland   thyroid gland   thyroxine   urea   urine

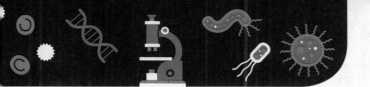
Maintaining water and nitrogen balance

Water leaves the body through the lungs during exhalation, and water, ions, and **urea** are lost from the skin in sweat. The body has no control over these losses.

Excess water, ions, and urea are removed by the kidneys in **urine**.

Levels of water in the body must be balanced because cells do not function efficiently if they lose or gain too much water.

The kidneys produce urine by filtration of the blood and selective reabsorption of useful substances such as water, glucose, and some ions.

The water level in the blood is controlled through this process by the hormone **ADH**, which affects the amount of water absorbed by the **kidney tubules**.

This is a negative feedback cycle.

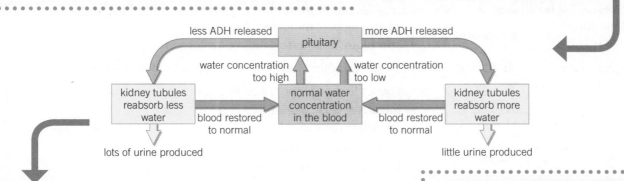

People who suffer from kidney failure may be treated by organ transplants or kidney **dialysis**.

Process of kidney dialysis

- blood temporarily removed from patient's body
- filtered through a dialysis machine
- patient's blood passes over dialysis fluid
- dialysis fluid has no urea
- urea and waste products diffuse from high concentration in patient's blood to low concentration in dialysis fluid
- patient's blood then returned to their body

Waste products

The digestion of proteins from food results in excess amino acids, which need to be excreted safely.

These amino acids are deaminated in the liver to form ammonia.

Ammonia is toxic, so it is immediately converted to urea for safe excretion.

Negative feedback

Negative feedback systems work to maintain a steady state. For example, blood glucose, water, and **thryoxine** levels are all controlled in the body by negative feedback.

Adrenaline

- produced by **adrenal glands** in times of fear or stress
- increases heart rate
- boosts delivery of oxygen and glucose to brain and muscles
- prepares the body for 'fight or flight' response
- does not involve negative feedback, as adrenal glands stop producing **adrenaline**

Thyroxine

- produced by the **thyroid gland**
- regulates how quickly your body uses energy and makes proteins (**metabolic rate**)
- important for growth and development
- levels controlled by negative feedback

B14 Hormonal coordination B

Hormones in human reproduction

During puberty, reproductive hormones cause the secondary sex characteristics to develop:

Oestrogen
- main female reproductive hormone
- produced in the **ovary**
- at puberty, eggs begin to mature and one is released every ~28 days

Testosterone
- main male reproductive hormone
- produced by the **testes**
- stimulates sperm production

Several hormones are involved in the **menstrual cycle**. Their functions are given in the table, and their levels vary as shown in the figure.

Hormone	Released by	Function
follicle stimulating hormone (FSH)	pituitary gland	• causes eggs to mature in the ovaries • stimulates ovaries to produce oestrogen
luteinising hormone (LH)	pituitary gland	• stimulates the release of mature eggs from the ovaries (**ovulation**)
oestrogen	ovaries	• causes lining of uterus wall to thicken • inhibits release of FSH • stimulates release of LH
progesterone	ovaries	• maintains thick uterus lining • inhibits release of FSH and LH

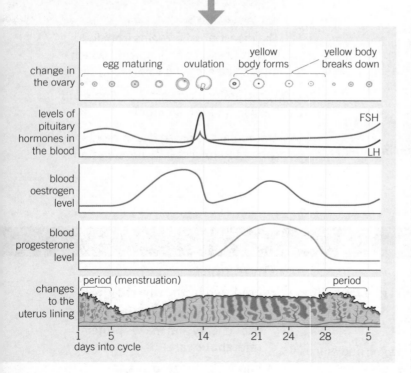

🐾 Revision tip

The names of the different hormones can be confusing, but it's really important that you use the correct names in the exam.

Contraception

Fertility can be controlled by a variety of hormonal and non-hormonal methods of **contraception**.

Hormonal contraception	Non-hormonal contraception
oral contraceptives – contain hormones to inhibit FSH production so no eggs mature	**barrier contraception**, for example, condoms and diaphragms – prevent sperm reaching the egg
	intrauterine devices – prevent the implantation of an embryo, or can release hormones like oral contraceptives
injection, **implant**, or skin patch – slow release of progesterone to inhibit maturation and release of eggs, which can last months or years	spermicidal agents – kill or disable sperm
	abstaining from intercourse when an egg may be in the oviduct
	surgical methods of male and female sterilisation

Treating infertility with hormones

Hormones are used in modern reproductive technologies to treat **infertility**.

FSH and LH can be given as a drug to treat infertility, or **in vitro fertilisation** (IVF) treatment may be used.

IVF treatment

1 mother given FSH and LH to stimulate the maturation of several eggs
2 eggs collected from the mother and fertilised by sperm from the father in a laboratory
3 fertilised eggs develop into embryos
4 one or two embryos are inserted into the mother's **uterus** (womb) when the embryos are still tiny balls of cells

Fertility treatment has some disadvantages:

- It is emotionally and physically stressful.
- It has a low success rate.
- It can lead to multiple births, which are a risk to both the babies and the mother.

Hormones in plants

Plants produce hormones to coordinate and control their growth, and their responses to light and gravity.

Phototropism is the orientation and growth of plants in response to light.

Geotropism (or **gravitropism**) is the growth of plants in response to gravity.

Uses of plant hormones

Plant hormones are used in agriculture and horticulture to control the growth of desirable plants and crops.

Plant hormone	Function	Agricultural uses
auxins	plant growth regulator – unequal distributions of auxin cause unequal growth rates in plant roots and shoots	• weedkiller • rooting powder • promoting growth in tissue cultures
ethene	acts as a hormone to control cell division	• control ripening of fruit during storage and transport
gibberellins	regulates developmental processes, including initiating seed germination	• end seed dormancy • promote flowering • increase fruit size

Key terms

Make sure you can write a definition for these key terms.

barrier contraceptive   contraception   geotropism   gravitropism   implant   infertility
intrauterine device   in vitro fertilisation   menstrual cycle   oral contraceptive   ovary
ovulation   phototropism   testes   uterus

Retrieval

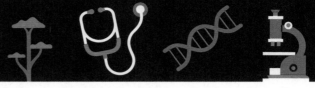

Learn the answers to the questions below then cover the answers column with
a piece of paper and write as many as you can. Check and repeat.

B14 questions	Answers
1 What is the endocrine system?	system of glands that secrete hormones into the bloodstream
2 How do the effects of the endocrine system compare to those of the nervous system?	endocrine system effects are slower but act for longer
3 Where is the pituitary gland located?	brain
4 Which organ monitors and controls blood glucose concentration?	pancreas
5 Which hormones interact to regulate blood glucose levels?	insulin and glucagon
6 What is the cause of Type 1 diabetes?	pancreas produces insufficient insulin
7 What is the cause of Type 2 diabetes?	body cells no longer respond to insulin
8 What is the function of the kidneys?	filter and reabsorb useful substances from the blood, and produce urine to excrete excess water, ions, and urea
9 How are excess amino acids excreted from the body?	deaminated to form ammonia in the liver, ammonia is converted to urea and excreted
10 Which hormone controls the water level in the body?	ADH
11 How is kidney failure treated?	organ transplant or kidney dialysis
12 What is the function of FSH?	causes eggs to mature in the ovaries, and stimulates ovaries to produce oestrogen
13 What is the function of LH?	stimulates the release of an egg
14 What is the function of oestrogen?	causes lining of uterus wall to thicken
15 What are the methods of hormonal contraception?	oral contraceptives, an injection, implant, or skin patch
16 State the disadvantages of IVF treatment.	• emotionally and physically stressful • low success rate • can lead to risky multiple births
17 What is the function of adrenaline in the body?	increases heart rate and boosts delivery of oxygen and glucose to brain and muscles to prepare the body for 'fight or flight'
18 What is the function of thyroxine in the body?	stimulates basal metabolic rate, so is important for growth and development
19 What is geotropism?	orientation and growth of plants in response to gravity
20 What is phototropism?	orientation and growth of plant in response to light
21 What are the uses of gibberellins in agriculture?	end seed dormancy, promote flowering, and increase fruit size

Put paper here

Now go back and use the questions below to check your knowledge from previous chapters.

Previous questions

Answers

Previous questions	Answers
Why are reflex actions described as rapid and automatic?	they do not involve the conscious part of the brain
What are the two parts of the central nervous system?	brain and spinal cord
Why are limiting factors important in the economics of growing plants in greenhouses?	greenhouses need to produce the maximum rate of photosynthesis while making profit
Give the limiting factors of photosynthesis.	temperature, carbon dioxide concentration, light intensity, and amount of chlorophyll
Write the balanced symbol equation for aerobic respiration.	$C_6H_{12}O_6 + 6O_2 \rightarrow 6CO_2 + 6H_2O$
Give the pathway of a nervous response.	stimulus $\rightarrow$ receptor $\rightarrow$ coordinator $\rightarrow$ effector $\rightarrow$ response
What is oxygen debt?	amount of oxygen the body needs after exercise to react with the accumulated lactic acid and remove it from cells

Put paper here

Required Practical Skills

Practise answering questions on the required practicals using the example below.
You need to be able to apply your skills and knowledge to other practicals too.

Plant responses	Worked example	Practice
For this practical, you will need to be able to make and record measurements of length and time to measure biological changes.	Two students measured the growth of plant seedlings in response to different light conditions. They grew five seedlings under partial light conditions, and five under full light conditions. Their heights were measured once a day for 20 days.	1 White mustard or cress seeds are often used to look at plant growth responses. Explain why these are an appropriate choice over acorns.
You will measure the length of newly-germinated seedlings from shoot to tip, then place groups of these seedlings in different levels of light and leave them to grow.	**1** Name the dependent variable in the investigation. Heights of the seedlings.	2 To achieve the partial light conditions the students placed the seedlings in boxes with a whole cut in the side. This caused the seedlings to grow sideways instead up upwards. Suggest how this could affect the results of the experiment.
Measuring the lengths of the seedlings over the a few days will tell you how much the plants grow in different light intensities. This practical also gives you the chance to practise making and labelling accurate biological drawings.	**2** Give three variables that the students should control. For example, the students should control: the amount of water given to all seedlings; the air temperature; air humidity; potting conditions (soil, cotton wool); measure both groups at the same time each day. **3** Which process causes the seedlings to grow upwards? Phototropism.	3 Explain why the heights of five seedlings were measured for each group.

 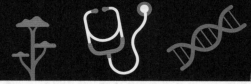

01 The graphs in **Figure 1** show changes in the levels of four hormones in a menstrual cycle.

Figure 1

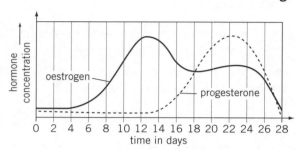 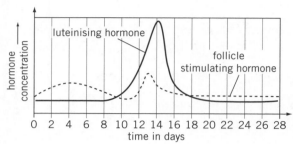

A teacher asked two students to analyse **Figure 1** and deduce on which day ovulation occurred.

One student thought that ovulation occurred on day 13, whereas the other student thought that ovulation occurred on day 14.

01.1 Use evidence from **Figure 1** and your own knowledge to explain which student was correct. **[4 marks]**

_____

_____

_____

_____

_____

> **! Exam Tip**
>
> With graphs that have more than one line on them it's important to be clear which set of data you're referring to.
>
> "The line on the graph shows…" won't be enough the get the marks.

01.2 Hormones are used in a number of contraceptives.

One type of contraceptive pill keeps the level of progesterone high for most of the cycle. This tablet has to be taken each day for 21 days of the menstrual cycle.

Use evidence from **Figure 1** to suggest how this might work. **[2 marks]**

prevent egg from being wrapped and released.

_____

01.3 Progesterone can also be released from a contraceptive implant placed under the skin of a woman's arm. Many women who have a contraceptive implant do not have a period during the time it is implanted.

Use evidence from **Figure 1** to suggest why their periods stop. **[2 marks]**

_____

_____

_____

> **! Exam Tip**
>
> You only need to use one line in **Figure 1** to answer this question.
>
> It helps to highlight the line you need so it stands out.

01.4 Evaluate the use of progesterone-releasing implants as a method of contraception. **[4 marks]**

_____

_____

_____

_____

_____

02 The endocrine system contains many glands.

02.1 Identify the glands marked **A–E** on **Figure 2**. **[5 marks]**

Figure 2

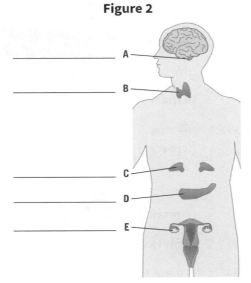

02.2 Explain why the pituitary gland is often referred to as a master gland. **[3 marks]**

_____

_____

_____

_____

02.3 Explain the role of the pituitary gland in maintaining the body's basal metabolic rate. **[6 marks]**

_____

_____

_____

_____

_____

_____

_____

03 Blood glucose levels in the body are constantly controlled.

03.1 Explain why blood glucose levels need to be maintained at a constant level. **[3 marks]**

03.2 People with diabetes have difficulty controlling their blood glucose levels. There are two main types of diabetes – Type 1 and Type 2. Describe the differences between Type 1 and Type 2 diabetes. **[4 marks]**

! **Exam Tip**

This question is just about the differences – don't waste time writing about similarities.

03.3 Compare the available treatments for Type 1 and Type 2 diabetes. **[4 marks]**

03.4 Suggest **two** actions that a government could take to reduce the number of new cases of diabetes. **[2 marks]**

04 Hormones play an important role in homeostasis.

04.1 Identify the hormone that is likely to increase after consuming a chocolate bar. Choose **one** answer. **[1 mark]**

adrenaline   insulin   glucagon   thyroxine

! **Exam Tip**

Start by crossing out any you know are wrong.

04.2 Explain your answer to **04.1**. **[4 marks]**

04.3 Many systems in homeostasis rely on negative feedback. Describe how a negative feedback system works. **[3 marks]**

05 Some couples have difficulty conceiving.

05.1 Explain **one** possible cause of infertility in females. **[2 marks]**

05.2 Explain **one** possible cause of infertility in males. **[2 marks]**

05.3 Some infertile couples receive in vitro fertilisation (IVF) treatment to increase their chances of getting pregnant. Describe the main steps involved in IVF treatment. **[4 marks]**

! **Exam Tip**

It's important to write a balanced argument for this question. Writing six advantages won't get you full marks.

05.4 Discuss the advantages and disadvantages of IVF treatment for those unable to conceive naturally. **[6 marks]**

05.5 **Figure 3** shows how the success rate for IVF varies with a female's age.

Figure 3

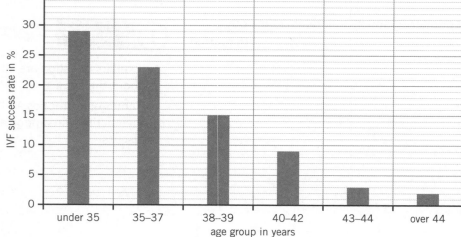

Each cycle of IVF treatment costs around £5000. In the UK qualifying women under the age of 40 are offered up to three IVF treatments without having to pay for it, via the National Health Service (NHS). Using **Figure 3**, evaluate the arguments for and against providing IVF treatment through the NHS. **[4 marks]**

06 **Figure 4** shows how the blood glucose concentration varies in a healthy person and in a person with Type 1 diabetes who treats themselves with insulin injections.

06.1 Describe why a person with Type 1 diabetes can't control their blood glucose concentration. **[1 mark]**

06.2 Compare the effect of eating a meal on a healthy person with the effect on a person who has Type 1 diabetes. **[2 marks]**

06.3 Using information from **Figure 4**, suggest **one** time at which the person with Type 1 diabetes injected insulin. **[1 mark]**

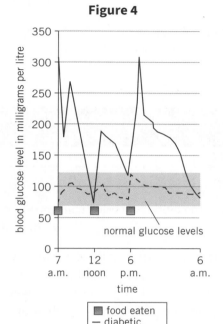

Figure 4

06.4 Calculate the percentage change in blood glucose concentration immediately following the midday meal for a person with Type 1 diabetes. **[2 marks]**

06.5 Discuss the potential for a cure for Type 1 diabetes. **[5 marks]**

07 Blood glucose levels are maintained in a healthy person by the action of insulin and glucagon.

07.1 Name the organ that produces these hormones. **[1 mark]**

07.2 Explain how blood glucose levels are maintained in a healthy person by the action of insulin and glucagon. **[6 marks]**

07.3 In 2015, approximately 3.5 million people in the UK were living with diabetes from a total population of 65 million. 9.4% of the US population has diabetes. Compare the rates of diabetes in the UK and the US. **[3 marks]**

08 The body responds to changes in its internal and external environment using the endocrine and nervous systems.

08.1 Write down the name given to a change that occurs in a person's environment. **[1 mark]**

08.2 Name the hormone that is released in response to a stressful situation. **[1 mark]**

08.3 Compare the actions of hormones in the endocrine system to nerves in the nervous system. **[6 marks]**

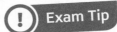
09 The kidney is responsible for removing some waste materials from the blood.

09.1 Name **one** waste product it removes. **[1 mark]**

09.2 **Table 1** shows the concentrations of plasma protein and glucose present in the blood entering the kidney, and in the urine it produces.

Table 1

Substance	Blood concentration in mg/ml	Urine concentration in mg/ml
plasma protein	700	0
glucose	100	0

Explain the differences between the blood and urine concentrations of protein and glucose. **[4 marks]**

09.3 The water content of urine produced throughout the day can vary. On a hot day, if a person does not drink a lot of water their urine contains very little water. Explain how the kidney controls the water content of the urine it produces on a hot day. **[4 marks]**

09.4 Diabetes can lead to kidney failure. Severe cases of kidney failure can lead to death. Explain **one** reason why severe kidney failure can lead to death. **[2 marks]**

09.5 Severe kidney failure can be treated using kidney dialysis or through a kidney transplant. Compare the advantages and disadvantages of kidney dialysis and kidney transplants. **[6 marks]**

10 Plant hormones have many practical uses.

10.1 Match the plant hormone to its use. **[2 marks]**

Plant hormone	Use of hormone
ethene	promote root growth in cuttings
gibberellins	control fruit ripening
auxins	increase fruit size

10.2 Auxins are also used as weedkillers. They cause uncontrolled growth of larger-leaved plants, causing them to die. Narrow-leaved plants, like wheat, are not affected in the same way. Explain why larger-leaved plants are more affected by auxin weedkillers than narrow-leaved plants. **[2 marks]**

10.3 Suggest and explain **one** advantage and **one** disadvantage of using auxin-based weedkillers when growing wheat. **[4 marks]**

11 A group of students were provided with three Petri dishes. Each dish contained 10 cress seeds that had germinated to produce plants that were approximately 5 cm tall. The students were also provided with the following apparatus:

- cardboard boxes with lids
- ruler
- scissors
- lamp
- forceps/tweezers

11.1 Describe how the students could use the apparatus to investigate the growth response of the cress seedlings when exposed to light from only one direction. **[4 marks]**

11.2 Predict and explain the results you would expect if the students removed the tips of three plants on each dish. **[6 marks]**

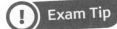

12 There is an old English saying that goes 'the child who picks a dandelion will wet their bed before the day is out'. Scientists have discovered that there is some truth in this superstition. Dandelion leaves contain a natural diuretic that increases urine production in the kidney. Suggest why eating dandelion leaves results in more urine being produced by the body. **[5 marks]**

13 Sucrose is an example of a carbohydrate molecule. It is made when two simple sugars, glucose and fructose, bind together.

13.1 Explain why sucrose is classified as a carbohydrate molecule. **[1 mark]**

13.2 Enzymes are used to make some soft-centred chocolates. An enzyme called invertase will catalyse the breakdown of sucrose into glucose and fructose. This causes chocolate to become softer and sweeter to the taste. Explain why invertase is an example of a catalyst. **[2 marks]**

13.3 Use the lock and key model to explain why invertase is only able to catalyse the breakdown of sucrose. **[3 marks]**

13.4 Suggest and explain **one** possible health benefit of adding invertase to chocolate. **[2 marks]**

14 Streptomycin is one type of antibiotic. It is used in the treatment of a number of infections, including tuberculosis (TB).

14.1 Identify the type of organism that causes TB. Choose **one** answer. **[1 mark]**

fungus   bacteria   virus   protozoa

14.2 Streptomycin works by inhibiting protein synthesis in the target pathogen. Name the cell component responsible for protein synthesis. **[1 mark]**

14.3 Suggest how streptomycin results in the death of the target pathogen. **[4 marks]**

B15 Variation

Variation in populations

Differences in the characteristics of individuals in a population are called **variation**.

Variation may be due to differences in

- the genes they have inherited, for example, eye colour (genetic causes)
- the environment in which they have developed, for example, language (environmental causes)
- a combination of genes and the environment.

Mutation

There is usually a lot of genetic variation within a population of a species – this variation arises from **mutations**.

A mutation is a change in a DNA sequence:

- mutations occur continuously
- very rarely a mutation will lead to a new phenotype, but some may change an existing phenotype and most have no effect
- if a new phenotype is suited to an environmental change, it can lead to a relatively rapid change in the species

Selective breeding

Selective breeding (artificial selection) is the process by which humans breed plants and animals for particular genetic characteristics.

Humans have been using selective breeding for thousands of years, since first breeding crops from wild plants and domesticating animals.

Process of selective breeding:
1 choose parents with the desired characteristic from a mixed population
2 breed them together
3 choose offspring with the desired characteristic and breed them
4 continue over many generations until all offspring show the desired characteristic

The characteristic targeted in selective breeding can be chosen for usefulness or appearance, for example:

- disease resistance in food crops
- animals that produce more meat or milk
- domestic dogs with a gentle nature
- larger or unusual flowers.

Disadvantages of selective breeding:

- can lead to **inbreeding**, where some breeds are particularly prone to inherited defects or diseases
- reduces variation, meaning all of a species could be susceptible to certain diseases

Cloning

A **clone** is an individual that has been produced **asexually** and is genetically identical to its parent. There are several different methods for producing both plant and animal clones, but there are benefits and risks associated with cloning.

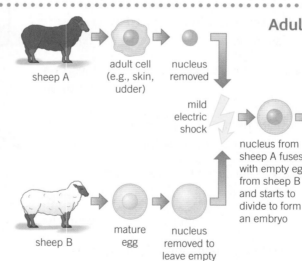

Adult cell cloning

sheep A → adult cell (e.g., skin, udder) → nucleus removed

mild electric shock

sheep B → mature egg → nucleus removed to leave empty egg

nucleus from sheep A fuses with empty egg from sheep B and starts to divide to form an embryo → the cloned embryo is implanted into the uterus of sheep C → lamb born is clone of sheep A, with the same genetic material

Key terms

Make sure you can write a definition for these key terms.

asexual   clone   cutting   embryo transplant   genetically modified   genetic engineering

inbreeding   mutation   selective breeding   tissue culture   variation

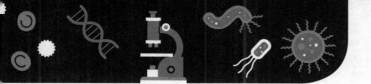

Methods of cloning

Tissue culture

Small groups of cells from part of a plant are used to grow identical new plants. This is important for preserving rare plant species and growing plants commercially in nurseries.

Cutting

An older, simple method used by gardeners to produce many identical plants from a parent plant.

Embryo transplant

Cells are split apart from a developing animal embryo before they become specialised, then the identical embryos are transplanted into host mothers.

Benefits	Risks
• large number of identical offspring produced • quick and economical • desired characteristics guaranteed	• limits variation and causes reduction in gene pool • clones may be vulnerable to diseases/changes in the environment • ethical considerations around cloning living organisms

Genetic engineering

Genetic engineering is a process that involves changing the genome of an organism by introducing a gene from another organism, to produce a desired characteristic.

For example:
- Bacterial cells have been genetically engineered to produce useful substances, such as human insulin to treat diabetes.
- Plant crops have been genetically engineered to be resistant to diseases, insects, or herbicides, or to produce bigger and better fruits and higher yields. Crops that have undergone genetic engineering are called **genetically modified** (GM).

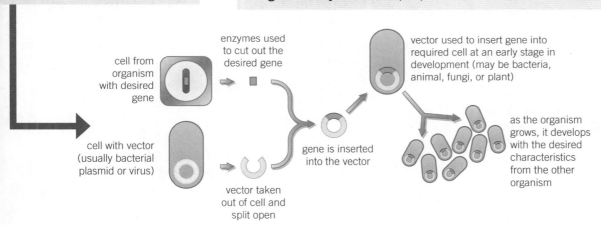

There are many benefits to genetic engineering in agriculture and medicine, but also some risks and moral objections.

Benefits	Risks
• potential to overcome some inherited human diseases • can lead to higher value of crops as GM crops have bigger yields than normal • crops can be engineered to be resistant to herbicides, make their own pesticides, or be more resistant to environmental conditions	• genes from GM plants and animals may spread to other wildlife, which could have devastating effects on ecosystems • potential negative impacts on populations of wild flowers and insects • ethical concerns, for example, in the future people could manipulate the genes of children to ensure certain characteristics • some believe the long-term effects on health of eating GM crops have not been fully explored

Learn the answers to the questions below then cover the answers column with a piece of paper and write as many as you can. Check and repeat.

B15 questions | Answers

1 What is variation?

differences in the characteristics of individuals in a population

2 What can cause variation?

genetic causes, environmental causes, and a combination of genes and the environment

3 How do new phenotype variants occur?

mutations

4 What is selective breeding?

breeding plants and animals for particular characteristics

5 Describe the process of selective breeding.

1 choose parents with the desired characteristic
2 breed them together
3 choose offspring with the desired characteristic and breed again
4 continue over many generations until all offspring show the desired characteristic

6 What are the consequences of inbreeding?

inherited defects and disease

7 What is genetic engineering?

modifying the genome of an organism by introducing a gene from another organism to give a desired characteristic

8 How have plant crops been genetically engineered?

to be resistant to diseases/herbicides/pesticides, to produce bigger fruits, to give higher yields

9 How have bacteria been genetically engineered?

to produce useful substances, such as human insulin to treat diabetes

10 What are enzymes used for in genetic engineering?

cut out the required gene

11 What is used to transfer the required gene into the new cell in genetic engineering?

vector (e.g., bacterial plasmid or virus)

12 Describe the steps involved in adult cell cloning.

1 nucleus removed from unfertilised egg cell
2 nucleus from adult body cell inserted into egg cell
3 electric shock stimulates egg cell to divide to form an embryo
4 embryo develops and is inserted into the womb of an adult female

13 What is tissue culture cloning?

using small groups of cells from plants to grow identical new plants

14 Why is tissue culture cloning of plants important?

preserve rare species and for growing plants commercially in nurseries

15 What is cutting as a cloning method?

simple method used by gardeners to produce many identical plants from a parent plant

16 Describe cloning through using embryo transplants.

cells split apart from a developing animal embryo before they are specialised, then the identical embryos are transplanted into host mothers

Put paper here

Now go back and use the questions below to check your knowledge from previous chapters.

Previous questions | Answers

Previous questions	Answers
What is the cause of Type 1 diabetes?	pancreas produces insufficient insulin
What are the methods of hormonal contraception?	oral contraceptives, an injection, implant, or skin patch
Give three things all control systems include.	receptors, coordination centres, and effectors
What do organisms need energy for?	• chemical reactions to build larger molecules • muscle contraction for movement • keeping warm
Give three internal conditions controlled in homeostasis.	body temperature, blood glucose concentration, and water levels
State the disadvantages of IVF treatment.	• emotionally and physically stressful • low success rate • can lead to risky multiple births

Put paper here

Maths Skills

Practise your maths skills using the worked example and practice questions below.

Mean, median, mode, and range

When interpreting data, scientists often need to calculate the average or range of a set of data.

You need to be able to calculate the range, and different types of averages (mean, median, and mode).

Mean: the calculated average of the numbers.

Median: the number in the middle.

Mode: the number that occurs most often.

Range: the difference between the largest and the smallest values.

Worked example

For the following list of numbers, we can calculate the mean, median, mode, and range.

13, 18, 13, 14, 13, 16, 14, 21, 13

Mean = 15

Add all the values together, and divide by the total number of values.

13 + 18 + 13 + 14 + 13 + 16 + 14 + 21 + 13 = 135

$\frac{135}{9} = 15$

Median = 14

Write the values in order and determine which one is in the middle. If is an even number of values, take the mean of the two values in the middle.

13, 13, 13, 13, [14] 14, 16, 18, 21

Mode = 13

13 appears the most times in the data.

Range = 8

The largest value (21) minus the smallest value (13).

21 − 13 = 8

Practice

Stomata on a leaf can be seen using a light microscope. The number of stomata differs depending on the location on the plant.

The table below gives the number of stomata a student counted from different areas on one leaf.

Leaf area	Number of stomata	
	Upper surface	Lower surface
1	4	42
2	1	43
3	0	45
4	4	43
5	1	39

Calculate the following:

1 The mean number of stomata on the lower surface of the leaf.

2 The median number of stomata on the upper surface of the leaf.

3 The mode number of stomata on the lower surface of the leaf.

4 The range in the number of stomata on the upper surface of the leaf.

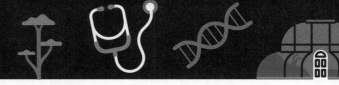
01 A market gardener grows and sells tomatoes. The gardener wishes to selectively breed the tomato plants so that the crops have a high yield and produce sweet tomatoes.

01.1 Write down the steps that the market gardener should take to produce a sweet-tasting, high-yielding variety of tomato plant. **[4 marks]**

> **! Exam Tip**
>
> This is an experiment on a larger scale. Think about how long steps might take and how many plants will be needed.

_____

_____

_____

_____

_____

_____

01.2 Suggest and explain **two** advantages to the market gardener of selectively breeding the tomato plants. **[4 marks]**

1 _____

_____

2 _____

_____

01.3 The gardener wishes to grow the tomato crop organically.

Suggest **one** other characteristic the gardener could selectively breed for to ensure a high yield of tomatoes is maintained. Give a reason for your answer. **[2 marks]**

> **! Exam Tip**
>
> If a plant is grown organically pesticides can't be used.

_____

_____

_____

02 A group of students investigated the variation in height that existed amongst students in their school.

Their results are shown in **Table 1**.

Table 1

Height in cm	$100 \leq h < 120$	$120 \leq h < 140$	$140 \leq h < 160$	$160 \leq h < 180$	$180 \leq h < 200$
number of students	12	18	36	22	6
midpoint					

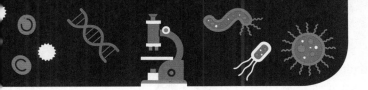

02.1 Plot an appropriate graph of these data. **[3 marks]**

Figure 1

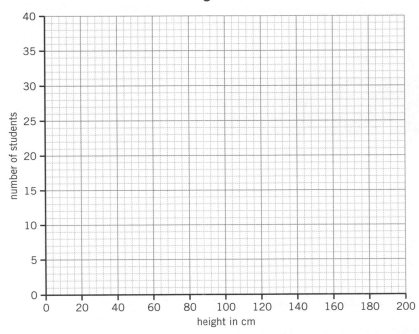

> **! Exam Tip**
>
> The choice of graph is important.
>
> Bar graphs can be used to show patterns in categoric data, line graphs can be used for continuous data, and histograms can be used to show patterns in frequency.

02.2 Explain the cause of the variation shown in the graph. **[2 marks]**

_____

_____

_____

02.3 The mean height of a person in the UK is 168 cm.

Calculate the mean height of the students in the school. **[4 marks]**

> **! Exam Tip**
>
> You may be used to using data tables to work out means in maths, but you may also be asked use this skill in science.

_____ cm

02.4 The students concluded that the distribution of heights in their school reflects that of the whole population.

Discuss the extent to which you agree, or disagree, with their conclusion. **[2 marks]**

_____

_____

_____

03 A group of scientists pooled their research data on differences caused by genetic and environmental factors. Three groups were studied:

- identical twins who were brought up together
- identical twins who were brought up separately
- non-identical twins who were brought up together

The data were summarised into the differences between the pairs studied, as shown in **Table 2**.

Table 2

	Mean difference in height in cm	Mean difference in mass in kg	Mean difference in IQ
identical twins, brought up together	1.4	1.1	4.9
identical twins, brought up separately	1.6	3.5	8.2
non-identical twins, brought up together	5.8	3.6	4.8

Each of the characteristics included in the study is affected by both a person's genes and their environment.

03.1 Identify the extent to which each factor is influenced by a person's genes or their environment. Justify your answers using data from **Table 2**. **[6 marks]**

_____

_____

_____

_____

_____

_____

_____

_____

03.2 Suggest **two** possible extensions to the study that would increase the validity of any conclusions made. **[2 marks]**

1 _____

_____

2 _____

_____

04 Some plant crops have been genetically modified to improve their characteristics.

04.1 Explain what is meant by the term genetically modified. **[2 marks]**

04.2 Reorder sentences **A–E** to describe how crops can be modified through genetic engineering. The first and last steps have been completed for you. **[4 marks]**

Step 1: The desired gene is removed from the nucleus of a donor cell.

A The bacteria are allowed to infect plant cells.

B The foreign gene is integrated into the plant cells' DNA.

C This 'foreign' gene is inserted into a plasmid (a circular piece of DNA).

D Bacteria reproduce quickly, producing many copies of the desired gene.

E The plasmid, now containing the desired gene, is inserted into a bacterial cell.

Step 7: Plant cells grow into plants displaying the desired characteristic.

 Exam Tip

Tick off each sentence as you use it, but don't cross it out completely in case you change your mind.

04.3 Give **one** advantage of the genetic modification of plant crops compared to selective breeding. **[1 mark]**

05 Most cattle and plant crops that are seen on farms today have been selectively bred.

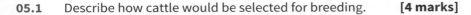

05.1 Describe how cattle would be selected for breeding. **[4 marks]**

05.2 Some crops are genetically engineered. Discuss the arguments for and against using genetic engineering to modify food crops. **[4 marks]**

 Exam Tip

Make sure your argument is balanced.

06 Insulin is a hormone secreted by the pancreas.

06.1 Explain the role of insulin in the body. **[2 marks]**

06.2 People with Type 1 diabetes have to inject themselves regularly with insulin. It is possible to genetically engineer bacteria so that they contain the gene for human insulin. This insulin can then be used to treat diabetes. Describe the main steps in the procedure of genetically engineering the bacteria this way. **[6 marks]**

06.3 Insulin originally used in the treatment of diabetes was extracted from pigs. Suggest **two** advantages of using insulin from genetically engineered bacteria rather than from pigs. **[2 marks]**

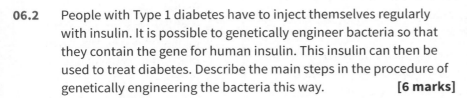 **Exam Tip**

'Describe' questions are asking for *what* happened, so go through the steps one at a time. Use key words and technical terms in the appropriate places.

07 Two students were asked to investigate whether plant leaf surface area is affected by light intensity. They decided to study the leaves on laurel bushes found in two locations: one area was in direct sunlight and the other was partially shaded by a building. They studied eight leaves in each section. The students recorded their results in **Table 3**.

Table 3

Section of laurel bush	Leaf surface area in cm²								Average
	1	2	3	4	5	6	7	8	
direct sunlight	72	72	68	72	68	68	70	70	70
shaded	72	76	70	64	82	72	84	72	

07.1 Complete **Table 3** by calculating the average leaf surface area for the laurel bushes found in the shade. **[1 mark]**

07.2 Calculate the uncertainty in leaf surface area for the shaded leaves. **[2 marks]**

07.3 One student concluded that leaves found in the shade have a larger surface area than leaves found in well-lit areas. The second student argued that it is not possible to form a conclusion from these data. Give **two** reasons why the second student was correct. Give reasons for your answers. **[4 marks]**

07.4 Laurel leaves have a complex shape. The students estimated the surface area of a leaf by drawing around the leaf on squared paper and adding up the number of squares contained within the shape. Suggest **one** alternative approach the students could have taken to estimate the leaf surface area. **[1 mark]**

08 A scientific magazine published the following article.

> *MioneTech, a US-based biotech company, reported today that it had successfully started the first human trials of a gene replacement therapy to cure Hunter syndrome, a previously incurable disorder. This inherited disorder prevents cells breaking down some sugars, leading to developmental delays, brain damage, and even death.*
>
> *The trial uses a form of DNA scissors called zinc finger nucleases (ZFNs), which cut both strands of the DNA double helix at a precise point. A virus is then used as the vector to transfer a 'healthy' replacement gene into the patient's DNA.*
>
> *The company claims that the revolutionary new treatment will change patients' lives for the better. People with Hunter syndrome require weekly infusions of a missing enzyme; the gene replacement therapy involves a single 3-hour operation.*

08.1 People with Hunter syndrome are missing a protein from their cells. Using the information from the article, name the type of protein molecule that the replacement gene will code for. **[1 mark]**

08.2 Evaluate the ethical and social issues of using gene replacement therapy to treat Hunter syndrome. **[6 marks]**

Exam Tip

Always look out for anomalous results when you're asked to calculate an average.

Exam Tip

Go thought the text with a set of highlighters.

Highlight the type of protein in blue, the social issues in pin, and the ethical issues in green.

Colour coding large blocks of text makes them easier to read and find the information you need.

09 Some species of tomato have been genetically engineered to be frost-resistant.

09.1 Suggest **two** advantages to a farmer of growing frost-resistant tomatoes. **[2 marks]**

09.2 To make frost-resistant tomatoes, genetic material is taken from a flounder fish. These flat fish are adapted to live in very cold water by producing an antifreeze chemical. Describe how frost-resistant tomatoes are created using genetic modification. **[6 marks]**

10 **Figure 2** shows a litter of puppies produced from the same parents.

Figure 2

10.1 The puppies show variation. Define the term variation. **[1 mark]**

10.2 Name **one** characteristic that is inherited. **[1 mark]**

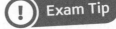

10.3 Give **one** characteristic that is affected by genes **and** the environment. Explain your answer. **[2 marks]**

10.4 Most breeds of dog have been selectively bred for certain characteristics. Give **one** disadvantage of selective breeding. **[1 mark]**

11.1 Define what is meant by a clone. **[1 mark]**

11.2 Identify which of the following techniques would **not** produce clones. Choose **one** answer. **[1 mark]**

tissue culture   selective breeding

plant cuttings   bacterial reproduction

11.3 Describe the main steps involved in producing new plants through tissue culture. **[4 marks]**

11.4 Give **one** advantage and **one** disadvantage of producing new plants through tissue culture. **[2 marks]**

12 Adult cell cloning is a relatively new technique.

12.1 Define what is meant by adult cell cloning. **[2 marks]**

12.2 Describe **one** possible use of adult cell cloning. **[1 mark]**

12.3 In 1997, Dolly the sheep was the first large mammal to be cloned. Describe the main steps involved in adult cell cloning used to produce Dolly. **[5 marks]**

! Exam Tip

While there are many possible uses, pick one that is likely to be on the mark scheme.

13 A group of students were asked to investigate the effect of temperature on the enzyme lipase. They chose to add lipase to full-fat milk and use an indicator to demonstrate when digestion had taken place. The students first mixed equal volumes of milk and sodium carbonate solution in a beaker. This caused the milk to become an alkaline solution. The students then used the following method:

1 Take 10 cm³ of the milk solution and add to a test tube. Place in an ice bath at 0 °C.

2 Leave the solution for 15 minutes to acclimatise.

3 Add a few drops of phenolphthalein indicator. Phenolphthalein is pink or purple in strongly alkaline solutions but becomes colourless when the pH drops below pH 8.

4 Start the stopwatch.

5 Shake the test tube to mix the contents thoroughly.

6 Stop the stopwatch when the solution becomes colourless.

7 Repeat steps 1–6, using water baths heated to 20 °C, 30 °C, 40 °C, 50 °C, and 60 °C.

! Exam Tip

Go through the method and pick out the variables.

This will give you a good understanding of what the student actually did.

13.1 Explain why the action of lipase on full-fat milk causes the phenolphthalein indicator to change colour from pink to colourless. **[3 marks]**

13.2 The students recorded their results in **Table 4**.

Table 4

Temperature in °C	Time in s	Reaction rate in s⁻¹
0	800	0.0013
20	170	
30	110	0.0091
40	110	
50	360	0.0028
60	–	–

! Exam Tip

s⁻¹ means 1/s
If you're not sure what to do, see if you can work out the reaction rates given.

Complete **Table 4** by filling in the missing values for the reaction rates at 20 °C and 40 °C. **[3 marks]**

13.3 Explain why the students were unable to gain a result at 60 °C. **[3 marks]**

13.4 Plot a graph of temperature versus rate of reaction to determine the optimum temperature for lipase action. **[3 marks]**

Figure 3

Exam Tip

Always draw a line of best fit. This can be straight or curved depending on the data.

13.5 Explain why the rate of reaction increased between 0 °C and 30 °C. **[3 marks]**

13.6 Suggest and explain **two** improvements the students could have made to their investigation. **[4 marks]**

14 Animal and plant cells both contain a number of sub-cellular structures. These include mitochondria and the nucleus.

14.1 Describe **two** other sub-cellular structures both types of cell contain. **[4 marks]**

14.2 The number of mitochondria varies between different cell types, as shown in **Table 5**.

Table 5

Human cell type	Mean number of mitochondria per cell (to the nearest 100)
small intestine	1600
skin	200
muscle	1700

Write down the range in the number of mitochondria in human cells. **[1 mark]**

Exam Tip

Think about the functions of the different cell types and what they need energy for.

14.3 Explain the data shown in **Table 5**. **[3 marks]**

Knowledge

B16 Reproduction A

Types of reproduction

<table>
<tr><th rowspan="6">Key facts</th><th></th><th>Sexual reproduction</th><th>Asexual reproduction</th></tr>
<tr><td>No. of parents</td><td>two parents</td><td>one parent</td></tr>
<tr><td>Cell division</td><td>cell division through **meiosis**</td><td>cell division through **mitosis**</td></tr>
<tr><td>Formulation</td><td>joining (fusion) of male and female sex cells (**gametes**) – sperm and egg in animals, pollen and ovule in plants</td><td>no fusion of gametes</td></tr>
<tr><td>Offspring</td><td>produces non-identical offspring that are genetically different to parents</td><td>produces offspring that are genetically identical to parent (**clones**)</td></tr>
<tr><td>Genetic variation</td><td>results in wide variation within offspring and species</td><td>no mixing of genetic information</td></tr>
<tr><td>Advantages</td><td></td><td><ul><li>produces variation in offspring</li><li>if the environment changes, the offspring may have a survival advantage by natural selection due to their genetic variation</li></ul></td><td><ul><li>only one parent needed</li><li>time and energy efficient as do not need to find a mate</li><li>faster than sexual reproduction</li><li>many identical offspring can be produced when conditions are favourable</li><li>successful traits passed on as offspring are identical</li></ul></td></tr>
<tr><td>Disadvantages</td><td></td><td><ul><li>finding a mate and reproducing is time consuming and requires lots of energy</li><li>much slower than asexual reproduction</li></ul></td><td><ul><li>reduced genetic variation – if the environment changes, the offspring may have a survival disadvantage</li><li>harmful mutations in parent would be passed on to all offspring</li></ul></td></tr>
</table>

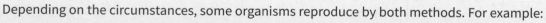

Depending on the circumstances, some organisms reproduce by both methods. For example:

- malaria parasites reproduce asexually in human hosts, but sexually in mosquitoes
- many fungi reproduce asexually by spores, but also sexually to give variation
- many plants produce seeds sexually, but also reproduce asexually by bulb division (daffodils) or runners (strawberry plants).

Key terms

Make sure you can write a definition for these key terms.

allele   chromosome   clone   dominant   fertilisation   gamete   gene

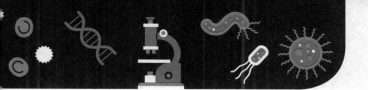

Meiosis

Meiosis is a type of cell division that makes gametes in the reproductive organs.

Meiosis halves the number of chromosomes in gametes, and **fertilisation** (joining of two gametes) restores the full number of chromosomes.

The fertilised cell divides by mitosis, producing more cells. As the embryo develops, the cells differentiate.

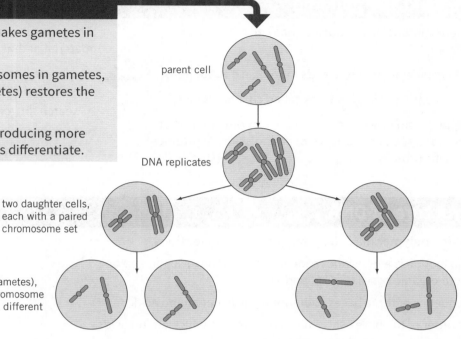

parent cell

DNA replicates

two daughter cells, each with a paired chromosome set

four daughter cells (gametes), each with a single chromosome set and all genetically different

Genetic inheritance

You need to be able to explain these terms about genetic inheritance:

gamete	specialised sex cell formed by meiosis
chromosome	long molecule found in the nucleus of cells made from DNA
gene	part of a chromosome that codes for a protein – some characteristics are controlled by a single gene (e.g., fur colour in mice and red-green colour-blindness in humans), but most are controlled by genes interacting
allele	different forms of the same gene
dominant	allele that only needs one copy to be expressed (it is always expressed)
recessive	allele that needs two copies present to be expressed
homozygous	when an individual carries two copies of the same allele for a trait
heterozygous	when an individual carries two different alleles for a trait
genotype	combination of alleles an individual has
phenotype	physical expression of the genotype – the characteristic shown

genotype   heterozygous   homozygous   meiosis   mitosis   phenotype   recessive

B16 Reproduction B

DNA and the genome

Genetic material in the nucleus of a cell is composed of **DNA**.

DNA is made up of two strands forming a **double helix**.

DNA is contained in structures called **chromosomes**.

A **gene** is a small section of DNA on a chromosome that codes for a specific sequence of amino acids, to produce a specific protein.

The **genome** of an organism is the entire genetic material of that organism.

The whole human genome has been studied, and this has allowed scientists to

- search for genes linked to different diseases
- understand and treat inherited disorders
- trace human migration patterns from the past.

Structure of DNA

DNA is a **polymer** made from four different **nucleotides**.

A nucleotide is a molecule made of a phosphate, a sugar, and one of four organic bases (A, C, G, and T).

A sequence of three bases codes for a particular amino acid.

The order of the bases determines the order in which amino acids are assembled to produce a specific protein.

In complementary DNA strands, a C base is always linked to a G base on the opposite strand, and a T to an A.

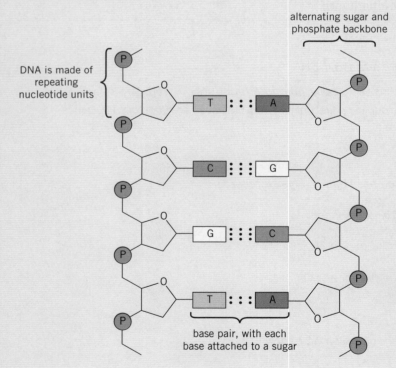

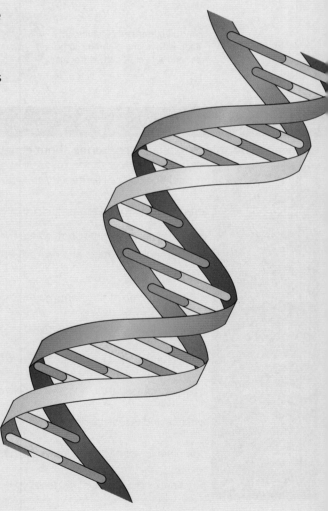

 **Key terms**

Make sure you can write a definition for these key terms.

chromosome   cystic fibrosis   DNA   double helix   expression   gene   genetic cross

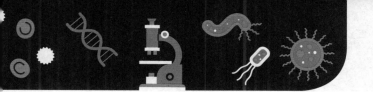

Protein synthesis

Proteins are synthesised on the **ribosomes** using a template of DNA.

Carrier molecules bring amino acids to add to the protein chain in the correct order.

When the protein is complete it folds up to form a specific shape, and this shape allows proteins to do specific jobs (as enzymes and hormones, or forming structures).

Non-coding parts of DNA can control the **expression** of genes by switching them on and off.

Inherited disorders

Some disorders are due to the inheritance of certain alleles:

- **Polydactyly** (extra fingers or toes) is caused by a dominant allele.
- **Cystic fibrosis** (a disorder of cell membranes) is caused by a recessive allele.

Embryo screening and gene therapy may alleviate suffering from these disorders, but there are ethical issues surrounding their use.

Mutations and genetic variability

Mutations occur continuously and change the base code of DNA. In coding DNA they may alter the activity of a protein:

- Most do not alter the appearance or function of the protein the DNA produces.
- A change in DNA structure may change the amino acid order, causing a gene to synthesise a different protein.
- Some mutations alter the shape of a protein, so the protein may no longer fit the substrate binding site, or lose its strength if it is structural.

In non-coding DNA, mutations may alter how genes are expressed.

Genetic crosses

A **genetic cross** is when you consider the offspring that might result from two known parents. **Punnett squares** can be used to predict the outcome of a genetic cross, for both the genotypes and phenotypes the offspring might have.

For example, the cross bb (brown fur) × BB (black fur) in mice:

		mother	
		B	B
father	b	Bb	Bb
	b	Bb	Bb

offspring genotype: 100% Bb

offspring phenotype: all black fur (B is dominant)

Sex determination

Normal human body cells contain 23 pairs of chromosomes – one of these pairs determines the sex of the offspring.

In human females the sex chromosomes are the same (XX, homozygous) and in males they are different (XY, heterozygous).

A Punnett square can be used to determine the probability of offspring being male or female. The probability is always 50% in humans as there are two XX outcomes and two XY outcomes.

		mother	
		X	X
father	X	XX	XX
	Y	XY	XY

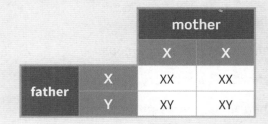

genome   mutation   nucleotide   polydactyly   polymer   Punnett square   ribosome

Learn the answers to the questions below then cover the answers column with a piece of paper and write as many as you can. Check and repeat.

	B16 questions		Answers
1	What is sexual reproduction?	Put paper here	joining (fusion) of male and female gametes
2	What type of cell division is involved in sexual reproduction?		meiosis
3	What type of cell division is involved in asexual reproduction?		mitosis
4	What is meiosis?	Put paper here	cell division that produces four daughter cells (gametes), each with a single set of chromosomes
5	What are the male and female sex chromosomes in humans?		XX – female, XY – male
6	How can plants produce asexually?	Put paper here	bulb division (e.g., daffodils) or runners (e.g., strawberry plants)
7	What is the genetic material in cells called?		DNA
8	What is DNA?		polymer made of chains of four different nucleotides
9	What does a nucleotide consist of?	Put paper here	sugar, phosphate group, and one of four different bases (A, G, C, or T)
10	What is the structure of DNA?		two complementary strands of nucleotides forming a double helix
11	What is a gene?		small section of DNA that codes for a particular amino acid sequence, to make a specific protein
12	How many bases code for an amino acid?	Put paper here	three
13	Which bases pair in complementary DNA strands?		C with G, T with A
14	What is the function of non-coding DNA?		switch genes on and off to control their expression
15	What are alleles?	Put paper here	different forms of the same gene
16	What is a recessive allele?		allele that needs to be present twice to be expressed
17	What is a dominant allele?		allele that is always expressed, even if only one copy is present
18	What is a genome?	Put paper here	the entire genetic material of an organism
19	Define the term homozygous.		two of the same alleles present in an organism
20	Define the term heterozygous.		two different alleles present in an organism
21	Where in the cell are proteins made?	Put paper here	on the ribosomes
22	What type of allele causes polydactyly?		dominant allele
23	What type of allele causes cystic fibrosis?		recessive allele
24	How many chromosomes do normal human body cells have?		23 pairs

Now go back and use the questions below to check your knowledge from previous chapters.

B16

Previous questions

Answers

Previous questions	Answers
What happens to muscles during long periods of activity?	muscles become fatigued and stop contracting efficiently
Describe the process of selective breeding.	1 choose parents with the desired characteristic 2 breed them together 3 choose offspring with the desired characteristic and breed again 4 continue over many generations until all offspring show the desired characteristic
Give the function of the cerebral cortex.	outer layer of the brain playing an important role in consciousness
What can cause variation?	genetic causes, environmental causes, and a combination of genes and the environment
What is the difference between aerobic and anaerobic respiration?	aerobic respiration uses oxygen, anaerobic respiration does not
How have bacteria been genetically engineered?	to produce useful substances, such as human insulin to treat diabetes

Put paper here

Maths Skills

Practise your maths skills using the worked example and practice questions below.

Probability	Worked Example	Practice
Probability is a number that tells you how likely something is to happen. It is important that you understand probability as this is key to genetic inheritance. For example, you could be asked to work out the probability of a child inheriting a genetic disease from its parents using a Punnett square. A value for probability can be expressed in the form of a fraction, decimal, or percentage. Probability can be calculated using the formula: $$\text{probability} = \frac{\text{number of ways the outcome can happen}}{\text{total number of outcomes}}$$	The Punnett square shows the inheritance of sex chromosomes in a genetic cross between two parents. <table><tr><td></td><td>X</td><td>Y</td></tr><tr><td>X</td><td>XX</td><td>XY</td></tr><tr><td>X</td><td>XX</td><td>XY</td></tr></table> male = XY female = XX What is the probability that the offspring from the genetic cross will be female? • number of ways the outcome can happen = 2 • total number of outcomes = 4 $$= \frac{2}{4} = 0.5$$ This probability can also be expressed as a fraction $\left(\frac{1}{2}\right)$ or a percentage (50%).	1 The Punnett square shows the inheritance of eye colour in a genetic cross. BB and Bb represent brown eyes. bb represents blue eyes. What is the probability that the offspring of the cross would have blue eyes? <table><tr><td></td><td>B</td><td>b</td></tr><tr><td>B</td><td>BB</td><td>Bb</td></tr><tr><td>b</td><td>Bb</td><td>bb</td></tr></table> 2 The Punnett square shows whether the offspring of a genetic cross between plants will be tall or short. TT and Tt represent tall plants, and tt represents short plants. What is the probability that the offspring of the cross would be tall? <table><tr><td></td><td>T</td><td>t</td></tr><tr><td>T</td><td>Tt</td><td>Tt</td></tr><tr><td>t</td><td>Tt</td><td>tt</td></tr></table>

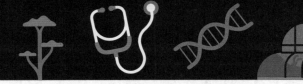
01.1 Complete the following sentences about genetic inheritance by circling the correct words. **[5 marks]**

Each human gamete contains **23 / 46 / 92** chromosomes. When a sperm and an egg fuse, a new cell called a zygote is formed. This new cell contains **23 / 46 / 92** chromosomes.

Two forms of each gene are inherited – these are called **alleles / dominant / recessive**.

One form of the gene is **allele / dominant / recessive** – this form is always expressed if present.

The other form of the gene is **allele / dominant / recessive** – a person must inherit this form of the gene from both parents if it is to be expressed.

> ! **Exam Tip**
>
> Take each sentence one at a time.
> Don't let the large block of text overwhelm you.

01.2 Eye colour is controlled by a gene. Two forms of the gene exist:
- brown eyes – dominant – **B**
- blue eyes – recessive – **b**

Classify each of the possible genotypes by matching the allele combination to its correct description. **[2 marks]**

Allele combination	Description
BB	homozygous recessive
Bb	homozygous dominant
bb	heterozygous

> ! **Exam Tip**
>
> There are lots of key words in this topic – make sure you're clear on all the different definitions.

01.3 Write down the phenotype for each of the allele combinations. **[3 marks]**

BB: _____

Bb: _____

bb: _____

01.4 A couple is expecting a baby. The father has blue eyes and the mother has brown eyes. Select the correct statement about their new baby's eye colour. **[1 mark]**

Tick **one** box.

The baby will definitely have brown eyes. ☐

The baby may be born with brown eyes or may be born with blue eyes. ☐

The baby will definitely have blue eyes. ☐

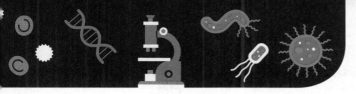

01.5 Give a reason for your answer to **01.4**. **[1 mark]**

_____

_____

02 **Figure 1** shows the sex chromosomes from two different people, **A** and **B**.

Figure 1

A   B

02.1 Identify which image represents the chromosomes from a female. Give reasons for your answer. **[2 marks]**

_____

_____

02.2 Carry out a genetic cross to show how sex is inherited. Use your diagram to show the likelihood of a couple having a baby girl. **[4 marks]**

Likelihood of having a baby girl: _____

02.3 A couple has three children. Calculate the probability that all three children are girls. **[3 marks]**

Probability: _____

02.4 A scientist studies two population groups:

- inhabitants of Manchester (population 500 000)
- inhabitants of Bath (population 90 000)

Explain which population is more likely to have a 1:1 ratio of males to females. **[3 marks]**

_____

_____

_____

03 Many people in the population have dimples. This is because having dimples is caused by a dominant allele **D**. The allele for no dimples is **d**.

03.1 Write down what is meant by a dominant allele. **[1 mark]**

_____

03.2 Which of the following allele combinations would lead to a child having dimples? Select as many combinations as required. **[1 mark]**

DD   **Dd**   **dd**

Exam Tip

There is a clue in the wording of the question. 'Select as many combinations as required' tells you it's probably more than one!

03.3 A couple decided to start a family. The alleles of the mother and father are shown in **Figure 2**. Complete the Punnett square to predict the possible allele combinations that the baby could have. **[2 marks]**

Figure 2

Mother's alleles	Father's alleles	
	D	d
D		
D		

Exam Tip

Approach Punnett squares logically – do either the rows first or the columns first.

03.4 Calculate and explain the probability of their child having dimples. **[2 marks]**

_____

_____

_____

03.5 If the parents have a second child, explain why the two children are likely to look similar but not the same. **[4 marks]**

_____

_____

_____

_____

_____

04 Polydactyly is an inherited condition. A couple is having a child; the father has polydactyly and is heterozygous for this disorder. The mother does not have polydactyly.

04.1 Polydactyly can be observed in a newborn baby. Write down the characteristic that would be observed for a baby who has inherited polydactyly. **[1 mark]**

_____

Exam Tip

Polydactyly is generally corrected by surgery during infancy, so the phenotype is not seen very often.

04.2 Explain how polydactyly is inherited. **[2 marks]**

_____

_____

04.3 Write down the allele combinations of the father and mother. **[2 marks]**

_____

_____

04.4 Over time, the mother and father have four more children. Draw a genetic cross diagram to show the possible alleles of the offspring. **[2 marks]**

04.5 Write down the expected ratio of children with polydactyly to children without polydactyly. **[1 mark]**

_____

04.6 Explain why the actual ratio of children with or without polydactyly is not necessarily the same as the ratio shown in **03.5**. **[3 marks]**

_____

_____

_____

_____

> (!) **Exam Tip**
>
> Think about the different options people have before having a child.

04.7 In the USA, 1 in 3500 babies are born with cystic fibrosis. In comparison, 1 in 500 babies are born with polydactyly. Suggest and explain **two** reasons for this trend. **[6 marks]**

_____

_____

_____

_____

_____

_____

_____

_____

05 Cystic fibrosis (CF) is an inherited disorder affecting over 10 400 people in the UK.

05.1 Write down **two** symptoms of CF. **[2 marks]**

05.2 Explain why a carrier (heterozygote) of CF will not suffer from the disorder. **[3 marks]**

> (!) **Exam Tip**
>
> This is an 'explain' question so use genetics to show *why*.

05.3 Draw a genetic diagram to show the possible genotypes of the offspring of two carriers of CF. **[2 marks]**

05.4 Using your diagram from **05.3**, calculate the percentage probability that a child born from two carriers of CF will inherit the disorder. **[2 marks]**

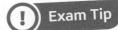

! Exam Tip

Even if you didn't get the correct answer for **05.4** you can still get the full marks for this question as an error in the previous question will be 'carried forward'.

This means you wouldn't be penalised for the same mistake twice.

05.5 Using the information below, and your answer to **05.4**, calculate the number of expected births in the UK with the genetic disorder CF in 2019.

- 1 in 25 people in the population are carriers of the CF allele
- expected number of births in 2019 is 700 000 **[5 marks]**

06 Sea anemones are animals that live in the oceans. They are closely related to corals. Sea anemones have the ability to reproduce sexually and asexually.

06.1 Name the type of cell division used in asexual reproduction. **[1 mark]**

! Exam Tip

Give a balanced argument – try to list an equal number of advantages and disadvantages.

06.2 Describe the main stages in this type of cell division. **[3 marks]**

06.3 Evaluate the advantages and disadvantages to an organism of being able to reproduce sexually and asexually. **[6 marks]**

07 Most species of tomato have 24 chromosomes present in the nucleus of their cells.

07.1 Write down how many chromosomes would be present in an adult tomato cell. **[1 mark]**

07.2 Write down how many chromosomes would be present in a tomato pollen cell. **[1 mark]**

! Exam Tip

Pollen cells are gametes.

07.3 Tomato plants reproduce sexually. Identify from the list below the **two** features that are present in sexual reproduction. **[2 marks]**

there is no mixing of genetic information   two parents are required

gametes fuse together   clones are produced

07.4 Describe the main steps in the production of a tomato pollen cell. **[4 marks]**

08 The genetic material found in the nucleus of human cells is composed of the chemical DNA.

08.1 Describe the main features in the structure of DNA. **[3 marks]**

08.2 Genes are small sections of DNA. Each gene contains a code. Describe what a gene codes for. **[2 marks]**

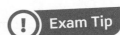

! Exam Tip

The question is only asking for benefits, so writing about disadvantages won't get you any marks.

08.3 In 2003, scientists announced that they had managed to sequence the entire human genome. Define the term genome. **[1 mark]**

08.4 Outline some of the important benefits of sequencing the human genome. **[3 marks]**

09 There are two types of cell division that occur in humans: meiosis and mitosis.

09.1 Write down where meiosis occurs in a female. **[1 mark]**

09.2 Write down where meiosis occurs in a male. **[1 mark]**

09.3 Compare the processes of mitosis and meiosis. **[4 marks]**

09.4 Explain why meiosis results in genetic variation. **[3 marks]**

09.5 Explain why the development of a fetus involves both mitosis and meiosis. **[4 marks]**

> **(!) Exam Tip**
>
> For a 'compare' answer give a statement about mitosis and then give the comparable statement for meiosis.

10 Cystic fibrosis (CF) is a genetically inherited disorder. People who have this condition produce excess mucus in their intestines. This can block digestive juices from the pancreas being released and can cover the villi.

10.1 Explain why people with CF are likely to be underweight if they do not follow a special diet. **[4 marks]**

10.2 Suggest why excess mucus production in the lungs leads to an increased rate of respiratory infections. **[2 marks]**

10.3 Currently there is no cure for CF. However, improved approaches to treatment have led to significant increases in life expectancy, from around 20 years old in 1980 to around 40 years old today.

A couple has a history of CF in both their families. After discovering they have conceived, the woman is offered an amniocentesis test. Amniocentesis involves taking fluid from around the developing fetus. The fluid contains fetal cells, which can be screened for the presence of the recessive allele that causes CF. The test is carried out at around 15 weeks of pregnancy. Discuss the social, economic, and ethical considerations involved in performing an amniocentesis test for this couple. **[6 marks]**

> **(!) Exam Tip**
>
> You are being testing to see if you can pull the important information out of a large block of text and mix it with what you have been taught:
>
> 1 read the text in full
> 2 go over the text and highlight all the social considerations in pink
> 3 highlight all the economic considerations in blue
> 4 highlight all the ethical consideration in green
> 5 make notes on any other considerations you can think of
> 6 write your answer, making sure it is balanced across the three points

11 The genetic material inside a cell is made of DNA. DNA is a polymer made of many nucleotides joined together.

11.1 Complete the following sentence about nucleotides.

Nucleotides consist of three main components: a base, a sugar,

and a _____. **[1 mark]**

11.2 The bases in DNA always bind together in the same pattern. Write down the sequence of bases for the complementary strand in **Figure 3**. **[2 marks]**

Figure 3

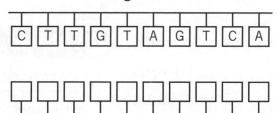

11.3 Explain how the bases in DNA lead to the formation of a specific protein. **[2 marks]**

11.4 Protein synthesis occurs at ribosomes. The protein keratin is the main component of hair. Describe the main steps in the synthesis of keratin. **[6 marks]**

12 Sickle cell anaemia is an inherited disorder that causes red blood cells to develop abnormally **(Figure 4)**.

Figure 4

normal
red blood cell

sickled
red blood cell

Sickle-shaped red blood cells cannot carry out their function properly.

12.1 Red blood cells are full of the protein haemoglobin. In a person with sickle cell anaemia haemoglobin does not function properly. Describe the function of haemoglobin. **[1 mark]**

12.2 Suggest **one** symptom of sickle cell anaemia. **[1 mark]**

12.3 Sickle cell anaemia is caused by a recessive allele. A mutation in the DNA led to a change in the order of one set of DNA bases. The bases GAG were changed to GTG. Explain how this DNA mutation can result in the production of a faulty version of haemoglobin. **[5 marks]**

12.4 Explain why not all base substitutions in DNA lead to changes in a person's phenotype. **[4 marks]**

13 An athlete volunteers to take part in a study of how the body responds to changes in its internal environment. On one day of the study, the athlete monitors their blood glucose levels hourly and the results are shown in **Figure 5**.

> ! **Exam Tip**
>
> Relate this to the function of red blood cells.

Figure 5

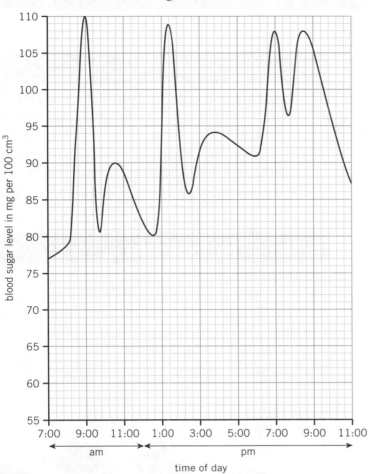

13.1 Give the name of the term that means 'to maintain a constant internal environment'. **[1 mark]**

13.2 Suggest and explain when the athlete ate lunch. **[2 marks]**

13.3 Explain the changes that took place in the athlete's body after eating lunch. **[4 marks]**

13.4 Calculate the percentage change in the athlete's blood sugar level between 7:00 am and 9:00 am. **[3 marks]**

13.5 A second volunteer in the test has Type 1 diabetes. Suggest how the graph would have looked different for this second volunteer, in comparison to the athlete's data. **[3 marks]**

14 Variation exists within all species.

14.1 Define the term variation. **[1 mark]**

14.2 You have two sunflower plants, **A** and **B**. You collect 20 seeds from sunflower **A** and 20 seeds from sunflower **B**. Using the seeds from the plants, suggest how you could carry out an investigation to determine the effect of genes and the environment on the growth of sunflower plants. **[6 marks]**

> **! Exam Tip**
> Use data from the graph to support your answer.

> **! Exam Tip**
> Draw lines on the graph to help you work out the values you need to use.

> **! Exam Tip**
> Plan this out step-by-step, clearly showing what you would do with the seeds and the results.

Knowledge

B17 Evolution A

Theory of evolution

Evolution is the gradual change in the inherited characteristics of a population over time.

Evolution occurs through the process of **natural selection** and may result in the formation of new species.

Darwin's work

Charles Darwin proposed the theory of evolution by natural selection after gathering evidence from a round-the-world expedition, experimentation, and discussion.

This states that all living species evolved from a common ancestor that first developed more than three billion years ago.

Darwin published this theory in *On the Origin of Species* (1859). His ideas were considered controversial and only gradually accepted because

- they challenged the idea that God made all of the Earth's animals and plants
- there was insufficient evidence at the time the theory was published, although much more evidence has been gathered since
- mechanisms of inheritance and variation were not known at the time
- other theories, such as that of Jean-Baptiste Lamarck, were based on the idea that the changes that occur in an organism over its lifetime could be passed on to its offspring. We now know that in the majority of cases this type of inheritance cannot occur.

Process of natural selection

The theory of evolution by natural selection states that

- organisms within species show a wide range of variation in phenotype
- individuals with characteristics most suited to the environment are more likely to survive and breed successfully
- these characteristics are then passed on to their offspring.

Evidence for evolution

The theory of evolution by natural selection is now widely accepted because there are lots of data to support it, such as

- it has been shown that characteristics are passed on to offspring in genes
- evidence from the **fossil record**
- the evolution of antibiotic resistance in bacteria.

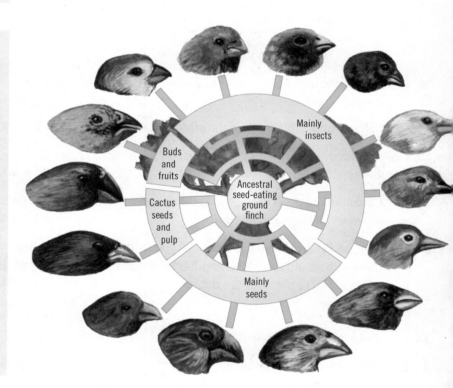

Key terms

Make sure you can write a definition for these key terms.

evolution   extinction   fossil record   natural selection   speciation

Speciation

Alfred Russel Wallace independently proposed the theory of evolution by natural selection.

He published joint writings with Darwin in 1858 on the subject, prompting Darwin to publish his book the next year.

Wallace worked worldwide gathering evidence for evolutionary theory.

He is best known for his work on warning colours in animals and for his pioneering work on the theory of **speciation**.

Speciation is the gradual formation of a new species as a result of evolution. More evidence and work from scientists over time have led to our current understanding of the theory of speciation.

Process of speciation

1 two populations of one species are isolated (e.g., by a river or mountain range)
2 natural selection occurs so that the better-adapted individuals reproduce and pass on these different characteristics
3 the populations have an increasing number of genetic mutations as they adapt to their different environments
4 eventually the two populations are so genetically different they cannot breed to produce fertile offspring

Extinction

Extinction is when there are no remaining individuals of a species still alive.

Factors that may contribute to a species' extinction include

- new predators
- new diseases or pathogens
- increased competition for resources or mates
- catastrophic events (e.g., asteroid impacts, volcanic eruptions, earthquakes)
- changes to the environment (climate change, destruction of habitats).

1 The reptile dies and falls to the ground

2 The flesh decays, leaving the skeleton to be covered in sand or soil and clay before it is damaged

3 Protected, over millions of years, the skeleton becomes mineralised and turns to rock. The rocks shift in the earth with the fossil trapped inside

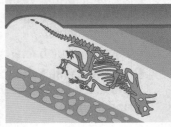
4 Eventually, the fossil emerges as the rocks move and erosion takes place

Fossils

Fossils are the remains of organisms from millions of years ago, which are found in rocks.

Fossils can be formed from

- parts of an organism that do not decay because one or more of the conditions needed for decay are absent
- hard parts of an organism (e.g., bones) when replaced by minerals
- preservation of the traces of organisms (e.g., burrows, footprints, and rootlet traces).

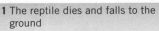

Benefits of the fossil record	Problems with the fossil record
- can tell scientists how individual species have changed over time - fossils allow us to understand how life developed over Earth's history - fossils can be used to track the movement of a species or its ancestors across the world	- many early organisms were soft-bodied, so most decayed before producing fossils - there are gaps in the fossil record as not all fossils have been found and others have been destroyed by geological or human activity – this means scientists cannot be certain about how life began on Earth

B17 Evolution B

Classification of living organisms

Carl Linnaeus developed a system to classify living things into groups based on their structure and characteristics.

New models of classification were proposed as understanding of biochemical processes developed and improvements in microscopes led to discoveries of internal structures.

There is now a **three-domain system** developed by Carl Woese, dividing organisms into

- Archaea (primitive bacteria usually living in extreme environments)
- Bacteria (true bacteria)
- Eukaryota (including protists, fungi, plants, and animals).

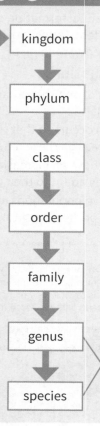

kingdom → phylum → class → order → family → genus → species

organisms are named by the **binomial system** of genus and species

Evolutionary trees

Evolutionary trees use current classification data for living organisms and fossil data for extinct organisms to show how scientists believe organisms are related.

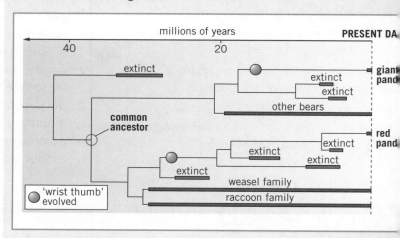

millions of years — PRESENT DAY

extinct · common ancestor · giant panda · extinct · extinct · other bears · red panda · extinct · extinct · extinct · extinct · weasel family · raccoon family · 'wrist thumb' evolved

Resistant bacteria

Bacteria can evolve rapidly because they reproduce very quickly.

This has led to many strains of bacteria developing **antibiotic resistance**, such as MRSA. The development of antibiotic resistance in bacteria is evidence for the theory of evolution by natural selection.

Emergence of antibiotic resistance

The development of new antibiotics is expensive and slow, so is unlikely to keep up with the emergence of new antibiotic-resistant bacteria strains.

To reduce the rise of antibiotic-resistant strains

- doctors should only prescribe antibiotics for serious bacterial infections
- patients should complete their courses of antibiotics so all bacteria are killed and none survive to form resistant strains
- the use of antibiotics in farming and agriculture should be restricted.

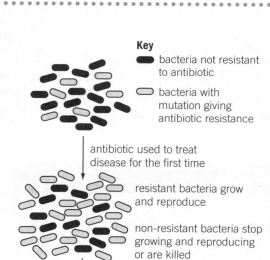

Key
- bacteria not resistant to antibiotic
- bacteria with mutation giving antibiotic resistance

antibiotic used to treat disease for the first time

resistant bacteria grow and reproduce

non-resistant bacteria stop growing and reproducing or are killed

antibiotic continues to be used

all bacteria now resistant to the antibiotic

selection has occurred for antibiotic resistance

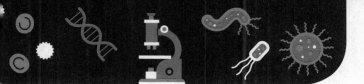

Understanding of genetics

Gregor Mendel developed our understanding of genetics by carrying out breeding experiments on plants in the mid-nineteenth century.

For example, he showed that crossing a plant that produces yellow peas and a plant that produces green peas always bred offspring with green peas. But when crossing these offspring, some offspring of later generations might have yellow peas again.

Through experiments like these, Mendel observed that the inheritance of each characteristic is determined by units – later called genes – that are passed on unchanged to offspring, and that these genes can be dominant or recessive.

The significance of Mendel's work was not recognised until after his death, because

- most scientists believed in blended inheritance (e.g., a white flower and a purple flower producing a lilac flower)
- he published his work in an obscure journal so not many people saw it
- he was a monk and not a scientist.

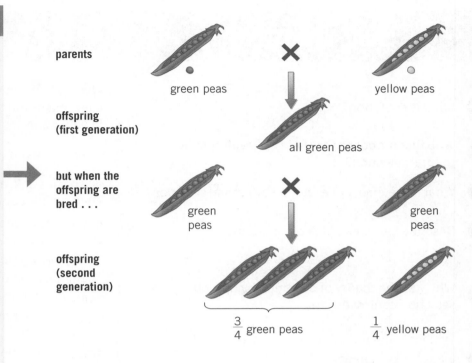

Development of gene theory

Further work by many scientists led to the development of **gene theory**.

> In the late nineteenth century the behaviour of chromosomes during cell division was observed.

> In the early twentieth century genes and chromosomes were observed to behave similarly, leading to the idea that genes were located on chromosomes.

> In the mid-twentieth century the structure of DNA and mechanism of gene function were determined.

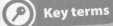 **Key terms**

Make sure you can write a definition for these key terms.

antibiotic resistance   binomial system   evolutionary tree

gene theory   three-domain system

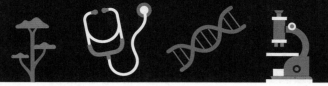

Learn the answers to the questions below, then cover the answers column with a piece of paper and write as many as you can. Check and repeat.

B17 questions	Answers
1 What is evolution?	change in the inherited characteristics of a population over time through natural selection, which may result in a new species
2 Who first proposed the theory of evolution by natural selection?	Charles Darwin
3 What is the theory of evolution by natural selection?	all species of living things evolved from a common ancestor that developed billions of years ago
4 Describe Lamarck's idea of inheritance.	organisms change over their lifetimes and these characteristics can be inherited
5 Why was the theory of evolution by natural selection controversial?	• challenged the idea that God made all of Earth's animals and plants • insufficient evidence at the time • genes, inheritance, and variation were not understood
6 What is speciation?	gradual formation of a new species as a result of evolution
7 What evidence supports the theory of evolution?	• parents pass on their characteristics to offspring in genes • fossil record evidence • evolution of antibiotic-resistant bacteria
8 What did Mendel discover through breeding experiments on plants?	inheritance of characteristics is determined by units (genes) passed on unchanged to offspring
9 What are fossils?	remains of organisms from millions of years ago, found in rocks
10 How might fossils be formed?	• parts of an organism do not decay because the conditions needed for decay are absent • traces of organisms are preserved • parts of an organism are replaced by minerals
11 What are the benefits of the fossil record?	can learn how species changed and life developed on Earth, and can track the movement of species across the world
12 What are the problems with the fossil record?	• many early organisms were soft-bodied so left few fossils • gaps in the fossil record as not all fossils have been found and some have been destroyed
13 What is extinction?	no individuals of a species are still alive
14 What is the binomial system?	naming of organisms by their genus and species
15 What classification system did Carl Woese introduce?	three-domain system of Archaea, Bacteria, and Eukaryota
16 Why can bacteria evolve rapidly?	they reproduce at a fast rate
17 How do antibiotic-resistant strains of bacteria develop?	mutations that allow the strain to survive and reproduce

Put paper here

Now go back and use the questions below to check your knowledge from previous chapters.

Previous questions

Answers

Previous questions		Answers
Which bases pair in complementary DNA strands?	Put paper here	C with G, T with A
How are monoclonal antibodies used to treat cancer?	Put paper here	for delivering toxic chemicals and drugs directly to cancer cells, limiting their harm to other cells in the body
Describe the steps involved in adult cell cloning.	Put paper here	**1** nucleus removed from unfertilised egg cell **2** nucleus from adult body cell inserted into egg cell **3** electric shock stimulates egg cell to divide to form an embryo **4** embryo develops and is inserted into the womb of an adult female
Name the four main components of blood.	Put paper here	red blood cells, white blood cells, plasma, platelets
What are artificial hearts used for?	Put paper here	keep patients alive while waiting for a transplant, or allow the heart to rest for recovery
What is a genome?		the entire genetic material of an organism
What are the male and female sex chromosomes in humans?	Put paper here	XX – female, XY – male
How many chromosomes do normal human body cells have?		23 pairs

Maths Skills

Practise your maths skills using the worked example and practice questions below.

Significant figures	Worked Example	Practice
Scientists often give numbers that are expressed to two or three significant figures (s.f.).	Zeros within a number count as significant figures. For example, 3.28034 has 6 significant figures.	**1** Round 0.009909 to 3 s.f.
	Leading zeros are never significant, so 0.00760 has 3 significant figures.	**2** Round 53879 to 2 s.f.
This is to avoid giving irrelevant or unnecessary figures in a very small or large number, or to avoid introducing error in a result.	**Example 1:**	**3** Round 0.005089 to 1 s.f.
	Round 2.837076 to 3 s.f.	**4** Round 98347 to 2 s.f.
	First count the significant figures from left to right, giving 2.83 to 3 s.f.	**5** Round 3.5175 to 3 s.f.
Significant figures can also be used to make large or complicated calculations easier.	As the 4th digit is a 7, the answer is rounded up, giving 2.84.	
	Example 2:	
A key point to remember is that leading zeroes (before a decimal point) are *never* significant.	Round 0.03601 to 3 s.f.	
	Number the significant figures, remembering that leading zeros are never significant.	
	As the 4th digit is a 1, the answer is not rounded up, giving 0.0360.	

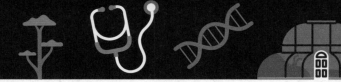

01 New species develop as a result of natural selection.

01.1 Define the term species. **[1 mark]**

_____

_____

01.2 **Figure 1** shows part of the evolutionary tree of primates.

Figure 1

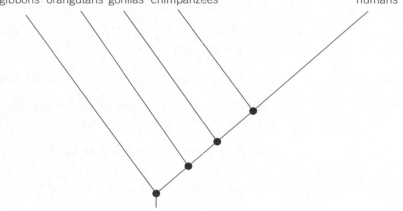

gibbons orangutans gorillas chimpanzees   humans

Identify the organism in **Figure 1** that most recently shared a common ancestor with humans. **[1 mark]**

_____

01.3 In 1858, Darwin proposed his theory of evolution by natural selection. Use the words from the box to complete the passage that describes how organisms evolve by natural selection. **[4 marks]**

characteristics	die	DNA	genotype
phenotype	similarities	survive	variation

Individual organisms within a species show _____ as a result of differences in their DNA. Individuals who have characteristics that are most suited to their environment are more

likely to _____ and reproduce. The individuals

pass on these favourable _____ to their offspring. This results in more individuals displaying these favourable

characteristics in their _____ .

> **Exam Tip**
>
> Not all the words will be used. Don't let this put worry you.

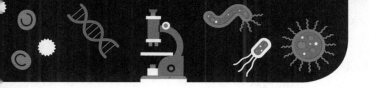

02 Some bacteria have evolved a resistance to antibiotics. One type of antibiotic-resistant bacteria is called MRSA. MRSA infections cause dizziness, nausea, high body temperature, and skin rashes. They can be fatal.

02.1 Explain how MRSA bacteria have evolved a resistance to antibiotics.

[4 marks]

_____

_____

_____

_____

_____

_____

> **! Exam Tip**
>
> Do not start your answer "MRSA bacteria have evolved a resistance to antibiotics because...". That is just repeating the question and will gain you no marks.

02.2 **Figure 2** shows the number of fatal cases of MRSA between 1995 and 2005 in the UK.

Figure 2

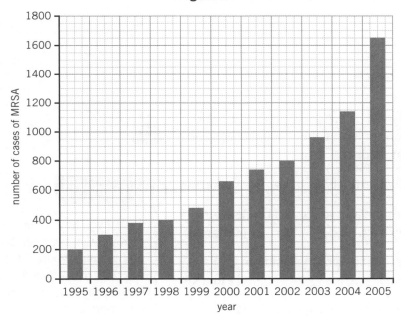

> **! Exam Tip**
>
> Draw lines across from the bar at 1995 and 2005 to help to read off the values on the *y*-axis.

Calculate the percentage change in the number of fatal cases of MRSA between 1995 and 2005. **[2 marks]**

_____ %

02.3 Suggest and explain **one** reason why the number of fatal cases of MRSA increased between 1995 and 2005. **[2 marks]**

_____

_____

_____

_____

02.4 The number of fatal cases of MRSA declined between 2006 and 2012. Suggest **two** reasons for this decline. **[2 marks]**

1 _____

_____

2 _____

_____

02.5 Antibiotics have been widely available since the mid-twentieth century. Explain why the evolution of antibiotic-resistant bacteria has occurred rapidly since then. **[2 marks]**

_____

_____

_____

_____

03 Dodos were flightless birds that lived on an island called Mauritius. When European sailors landed on the island, they brought dogs and rats with them. The dodo became extinct.

03.1 Describe what is meant by the term extinction. **[1 mark]**

03.2 Using the information in the question and your own knowledge, suggest why the dodo became extinct. **[4 marks]**

03.3 In 2016, the dodo genome was sequenced for the first time. Explain what is meant by the term genome. **[1 mark]**

03.4 Suggest **two** reasons for sequencing the dodo genome. **[2 marks]**

! **Exam Tip**

Even if you have no knowledge of your own, the main body of the question provides you with lots of useful information, such as _flightless birds_ and _dogs_ and _rats_.

04 Water birds have long, webbed feet. One theory of evolution suggests that over the years, due to straining their toes to paddle through the water, these birds developed elongated, webbed toes that help them swim more efficiently.

04.1 Name the scientist that proposed this theory of evolution. **[1 mark]**

04.2 The scientist suggested that webbed feet evolved as a result of natural selection. Describe how webbed feet would evolve by natural selection. **[4 marks]**

! **Exam Tip**

Remember to relate you answer to change over years. This was referred to in the main part of the question.

04.3 Give **one** reason why people did not believe Darwin's theory when he first proposed it. **[1 mark]**

04.4 Darwin's theory of evolution by natural selection is now widely accepted. Give **one** piece of evidence for evolution by natural selection. **[1 mark]**

05 A group of scientists develop a drug called 'Drug 2030' to treat bacterial infections. In trials, the drug proves highly successful at treating many bacterial infections. However, some bacteria have a natural resistance to the active agent in Drug 2030.

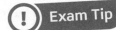

(!) Exam Tip

This is a hypothetical drug to see if you can apply what you know about other drugs and developing antibiotic resistance to a new situation. Do not spend any time worry about what this drug does, or panicking that you haven't heard of this before.

05.1 Explain how resistance to Drug 2030 could lead to changes in the population of bacteria. **[6 marks]**

05.2 Describe the features of a successful antibiotic drug. **[2 marks]**

05.3 An increasing number of strains of bacteria are becoming resistant to antibiotics. Explain **two** steps that people should follow to help reduce the number of resistant strains of bacteria. **[4 marks]**

06 Scientists have sorted all known living organisms into groups.

06.1 Describe how organisms are classified into groups according to the Linnaean classification system. **[2 marks]**

06.2 Blue tits are common garden birds. The blue tit's scientific name is *Cyanistes caeruleus*. Identify which **two** of the following statements are true about blue tits. **[2 marks]**

The blue tit belongs to the kingdom Animalia.

The blue tit belongs to the species Chordata.

The blue tit belongs to the phylum *caeruleus*.

The blue tit belongs to the genus *Cyanistes*.

06.3 A number of different classification systems have been proposed. Explain **two** advances that have taken place that have led to the development of new classification systems. **[4 marks]**

07 Ammonite fossils are often found on beaches in the UK. Ammonites were marine animals belonging to the phylum Mollusca. They had a coiled external shell which provided protection. Ammonites probably fed on small plankton or vegetation growing on the sea floor.

07.1 Explain why only the shell of the animal is seen in an ammonite fossil. **[3 marks]**

07.2 The ammonites became extinct at the end of the Cretaceous period, at roughly the same time as the dinosaurs disappeared.

Suggest and explain **two** reasons why ammonites became extinct.

[4 marks]

07.3 The fossil record shows that ammonites lived between approximately 200 million and 60 million years ago with little change to their basic structure. Explain why these organisms did not evolve significantly over this period of time. **[2 marks]**

08 Carl Woese developed a three-domain system of classification.

08.1 Which of the following is a domain in Woese's classification system? Choose **one** answer. **[1 mark]**

Vertebrates   Archaea   Chordata   *Felis*

08.2 Give the name of the domain that dogs belong to. **[1 mark]**

08.3 Describe how DNA analysis helped Woese develop his classification system. **[4 marks]**

08.4 Evolutionary trees are built up through DNA analysis, similarities in characteristics, and fossil evidence. **Figure 3** shows an example of an evolutionary tree.

Figure 3

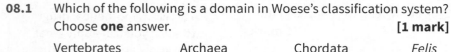

> **! Exam Tip**
>
> Give evidence from the tree, just as you would from a graph.

Give **three** conclusions that can be drawn from the evolutionary tree in **Figure 3**. **[3 marks]**

09 Speciation is the formation of new species in the course of evolution.

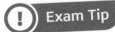

09.1 Which of the following statements is true about speciation? Choose **one** answer. **[1 mark]**

A new species has formed when the new organism cannot produce fertile offspring when breeding with the ancestor species.

A new species has formed when the new organism can produce fertile offspring when breeding with the ancestor species.

A new species has formed when the new organisms can produce fertile offspring with each other.

A new species has formed when the new organisms have different behaviours to the ancestor species.

> **! Exam Tip**
>
> Put a cross next to the ones you know are incorrect and a question mark by the ones you are not sure of.

09.2 Which scientist developed the theory of speciation by studying warning colouration in animals? Choose **one** answer. **[1 mark]**

Darwin   Wallace   Lamarck   Mendel

09.3 Explain **one** way in which speciation can occur. **[3 marks]**

10 Life on Earth began 4 billion years ago. It is estimated that 5 billion different species have lived on Earth. Of these, it is estimated that 14 million species are alive today, and 1.2 million species alive today have been identified and classified.

10.1 Calculate the proportion of species estimated to be alive that have not yet been identified. **[3 marks]**

10.2 Suggest **two** reasons why the proportion of species identified is only a small fraction of the total estimated number of species. **[2 marks]**

! Exam Tip

Be careful with which numbers you select for the calculation in **10.1**.

10.3 Explain why over 99 % of species that ever lived can no longer be found alive on Earth. **[3 marks]**

10.4 It is estimated that 1250 species per year become extinct. Compare the current rate of extinction with the mean extinction rate since life began on Earth. **[4 marks]**

10.5 Suggest **three** reasons for the difference in extinction rates calculated in **10.4**. **[3 marks]**

11 Overfishing in the North Sea between the 1960s and 1990s significantly depleted fish stocks. To try to support fish populations the Government brought in a number of new rules for deep sea fishing. One of these rules requires the use of large-mesh fishing nets that contain large holes.

11.1 Suggest and explain why fishing vessels were required to use nets with large holes. **[3 marks]**

11.2 Following the period of heavy fishing in the North Sea, scientists discovered that fish species had started to breed even when they were still small.

Explain how changes to the minimum fish breeding size occurred. **[6 marks]**

11.3 **Figure 4** shows the estimated North Sea cod population from 1970 to 2015.

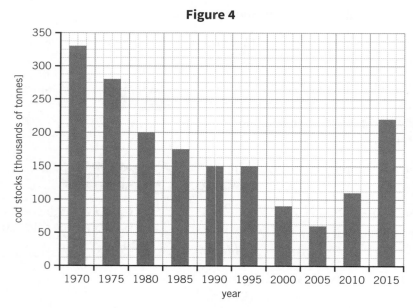

Figure 4

> **!** **Exam Tip**
>
> The *y*-axis is in thousands of tonnes. Be sure to use this unit in you answer. Just using tonnes would be incorrect.

Evaluate the success of the Government's strategy to support North Sea fish stocks. **[3 marks]**

11.4 Estimate the North Sea cod stocks for the year 2020. Justify your answer. **[4 marks]**

12 In the mid-nineteenth century, a series of experiments on pea plants developed our understanding of genetics.

12.1 Name the scientist who carried out these experiments. **[1 mark]**

12.2 One of the experiments investigated crossing green pea plants with yellow pea plants. Alleles for green-coloured peas (G) are dominant. Alleles for yellow-coloured peas (g) are recessive.

Use the Punnett square diagram to calculate the likelihood of two heterozygous green pea plants producing offspring with yellow peas. **[2 marks]**

> **!** **Exam Tip**
>
> Do these one or two columns at a time. One place to start is with parent plant A and allele G, then parent plant A allele g, and so on. This way you're less likely to get confused and repeat one allele or miss one allele out.

		Parent plant A	
		G	g
Parent plant B	G		
	g		

12.3 Use the most appropriate bold words to complete the sentences about the pea plant experiments. **[4 marks]**

The scientist noticed that characteristics are inherited in a clear, **random** / **predictable** pattern. He explained his experiments by suggesting that organisms contain different units of **inherited** / **environmental** material which can be passed on to offspring. The units never mix, and some characteristics are **dominant** / **recessive** over others. These units of inheritance are now called **genes** / **cells**.

Exam Tip

Take this one sentence at a time. It you can't decide between random or predictable at the start, don't let this prevent you from trying the second sentence.

12.4 Suggest **one** reason why the work was not accepted by the scientific community at the time. **[1 mark]**

13 People with anaemia often have a low red blood cell count.

13.1 Suggest **one** symptom of anaemia. **[1 mark]**

13.2 Explain how a red blood cell is adapted to its function. **[3 marks]**

13.3 Erythropoietin (EPO) is a drug that can be used to treat people with severe anaemia. It increases the number of red blood cells in the bloodstream. Athletes are banned from using this drug.

Suggest and explain why athletes are banned from using EPO. **[4 marks]**

14 Some students set up an experiment to investigate osmosis. They set up the equipment shown in **Figure 5** and left it for 15 minutes.

Figure 5

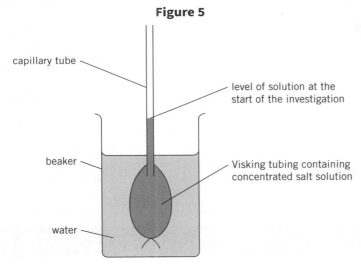

14.1 Complete the sentence with the correct bold word. **[1 mark]**

The salt solution is **isotonic** / **hypertonic** / **hypotonic** to the water.

14.2 Predict and explain the results you would expect the students to find. **[3 marks]**

14.3 Explain what would happen if a cell taken from a leaf was placed into the solution inside the Visking tubing. **[4 marks]**

B18 Adaptation

Ecosystem organisation

> **Individual organisms**

↓

> **Population**
> the total number of organisms of the same species that live in one specific geographical area

↓

> **Community**
> group of two or more populations of different species living in one specific geographical area

↓

> **Ecosystem**
> the interaction of a community of living organisms with the non-living parts of their environment

A stable community is one where all the species and environmental factors are in balance so that population sizes remain fairly constant.

An example of this is the interaction between predator and prey species, which rise and fall in a constant cycle so that each remains within a stable range.

Competition

To survive and reproduce, organisms require a supply of resources from their surroundings and from the other living organisms there.

This can create competition, where organisms within a community compete for resources.

There are two types of competition – **interspecific competition** is between organisms of different species and **intraspecific competition** is between organisms of the same species.

Animals often compete for
- food
- mates
- territory.

Plants often compete for
- light
- space
- water and mineral ions from the soil.

Interdependence

Within a community each species **interacts** with many others and may depend on other species for things like food, shelter, pollination, and seed dispersal.

If one species is removed it can affect the whole community – this is called **interdependence**.

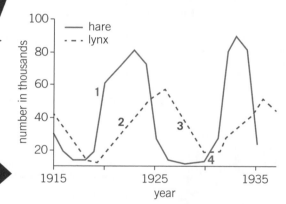

1 If the population of hares increases there is a larger food supply for the lynx.

2 This can therefore support m[ore] lynx, so more offspring survive.

3 The growing numbers of lynx eventually reduce the food sup[ply]. The number of predators starts [to] decrease.

4 The prey population starts to increase once more – the cycle then begins again.

Abiotic factors

Abiotic factors are non-living factors in the ecosystem that can affect a community.

Too much or too little of the following abiotic factors can negatively affect the community in an ecosystem:
- carbon dioxide levels for plants
- light intensity
- moisture levels
- oxygen levels for animals that live in water
- soil pH and mineral content
- temperature
- wind intensity and direction

Biotic factors

Biotic factors are living factors in the ecosystem that can affect a community.

For example, the following biotic factors would all negatively affect populations in a community:
- decreased availability of food
- new predators arriving
- new pathogens
- competition between species, for example, one species outcompeting another for food or shelter, causing a decline in the other species' population

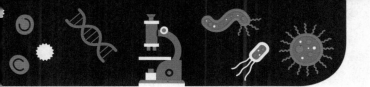
Adaptations of organisms

Organisms have features – **adaptations** – that enable them to survive in the conditions in which they live. The adaptations of an organism may allow it to outcompete others, and provide it with an evolutionary advantage.

Structural adaptations

The physical features that allow an organism to successfully compete:

- sharp teeth to hunt prey
- colouring that may provide camouflage to hide from predators or hunt prey
- a large or small body-surface-area-to-volume ratio

Behavioural adaptations

The behaviour of an organism that gives it an advantage:

- making nests to attract a mate
- courtship dances to attract a mate
- use of tools to obtain food
- working together in packs

Functional adaptations

Adaptations related to processes that allow an organism to survive:

- photosynthesis in plants
- production of poisons or venom to deter predators and kill prey
- changes in reproduction timings

You can work out how an organism is adapted to where it lives when given information on its environment and what it looks like.

For example, without the following adaptations the organisms below would be at a disadvantage in their environment.

Organism	Example adaptations
	<ul><li>white fur for camouflage when hunting prey</li><li>feet with large surface area to distribute weight on snow</li><li>small ears to reduce heat loss</li><li>thick fur for insulation</li></ul>
	<ul><li>feet with large surface area to distribute weight on sand</li><li>hump stores fat to provide energy when food is scarce</li><li>tough mouth and tongue to allow camel to eat cacti</li><li>long eyelashes to keep sand out of eyes</li></ul>
	<ul><li>spines instead of leaves to reduce surface area and therefore water loss, and to deter predators</li><li>long roots to reach water underground</li><li>large, fleshy stem to store water</li></ul>

Some organisms are **extremophiles**, which means they live in environments that are very extreme where most other organisms could not survive. For example, areas with

- very high or low temperatures
- extreme pressures
- high salt concentrations
- highly acidic or alkaline conditions
- low levels of oxygen or water.

Bacteria that live in deep sea vents are extremophiles.

Deep sea vents are formed when seawater circulates through hot volcanic rocks on the seafloor. These environments have very high pressures and temperatures, no sunlight, and are strongly acidic.

 **Key terms**

Make sure you can write a definition for these key terms.

abiotic factor   adaptation   biotic factor   community   ecosystem   extremophile
interaction   interdependence   interspecific competition   intraspecific competition   population

Learn the answers to the questions below, then cover the answers column with
a piece of paper and write as many as you can. Check and repeat.

B18 questions	Answers
1 What is a population?	total number of organisms of the same species that live in a specific geographical area
2 What is a community?	group of two or more populations of different species living in a specific geographical area
3 What is an ecosystem?	the interaction of a community of living organisms with the non-living parts of their environment
4 What is competition?	contest between organisms within a community for resources
5 What is interdependence?	when species in a community depend on others for resources and shelter
6 What do animals often compete for?	food, mates, and territory
7 What do plants often compete for?	light, space, water, and mineral ions
8 What is an abiotic factor?	non-living factor that can affect a community
9 List the abiotic factors that can affect a community.	• carbon dioxide levels for plants • light intensity • moisture levels • oxygen levels for animals that live in water • soil pH and mineral content • temperature • wind intensity and direction
10 What is a biotic factor?	living factor that can affect a community
11 List the biotic factors that can affect a community.	• availability of food • new predators • new pathogens • competition between species
12 What is a stable community?	when all species and environmental factors are in balance, so population sizes remain fairly constant
13 How do adaptations help an organism?	they enable the organism to survive in the conditions in which it lives
14 What are the three types of adaptations?	structural, behavioural, and functional
15 What is an extremophile?	an organism that lives in a very extreme environment
16 What makes an environment extreme?	• very high or low temperatures • extreme pressures • high salt concentrations • highly acidic or alkaline conditions • lack of oxygen or water

Put paper here

Now go back and use the questions below to check your knowledge from previous chapters.

B18

Previous questions

Answers

Question		Answer
What classification system did Carl Woese introduce?	Put paper here	three-domain system of Archaea, Bacteria, and Eukaryota
How might fossils be formed?	Put paper here	• parts of an organism do not decay because the conditions needed for decay are absent • traces of organisms are preserved • parts of an organism are replaced by minerals
Describe cloning through using embryo transplants.	Put paper here	cells split apart from a developing animal embryo before they are specialised, then the identical embryos are transplanted into host mothers
Give two common defects of the eyes.	Put paper here	myopia (short-sightedness) and hyperopia (long-sightedness)
Describe how temperature affects the rate of photosynthesis.	Put paper here	increasing temperature increases the rate of photosynthesis as the reaction rate increases – at high temperatures enzymes are denatured so the rate of photosynthesis quickly decreases
Why are antibodies a specific defence?		antibodies have to be the right shape for a pathogen's unique antigens, so they target a specific pathogen
What is the function of the nervous system?		It enables organisms to react to their surroundings and coordinates behaviour

Maths Skills

Practise your maths skills using the worked example and practice questions below.

Estimations	Worked Example	Practice
Estimates are often used in science before an exact calculation is done, such as when dealing with large numbers like population sizes. For example, if you work out how many snails live in $1\,m^2$, you can use this value to estimate how many snails live in an area of $10\,m^2$. You can also make estimates from sets of data. To make an estimate based on a graph, try drawing a line or curve of best fit through the data points. This will enable you to draw a straight line tangent between two points, from which you can make an estimate.	A grassy field on a farm measured 120 metres by 90 metres. A student wanted to estimate the number of daisies growing in the field. The student placed a $1\,m \times 1\,m$ quadrat in one position in an area that daisies were found in.  quadrat To estimate the number of buttercup plants in the field: Number of buttercups in $1 \times 1\,m$ quadrat = 7 Area of the field (120×90) = $10800\,m^2$ $7 \times 10800 = 75\,600$ daisies estimated in the field	1 The average number of dandelions in a 1 metre by 1 metre square of a park is 6. The park measures 230 metres by 350 metres. Estimate the number of dandelions in the park. 2 The average number of daisies in a 1 metre by 1 metre square of a field is 28. The field measures 180 metres by 80 metres. Estimate the number of daisies in the field.

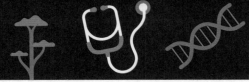
Exam-style questions

01 Over the past 50 years, scientists in Italy have been monitoring the populations of red and grey squirrels. Red squirrels are native to the Piedmont region; grey squirrels were introduced to the area in the mid 20th century.

To monitor the populations of these species, the scientists divided a map of the area into squares of equal size, known as cells. They then recorded the number of cells in which each species of squirrel was present.

01.1 Evaluate whether this technique provides data on population size.
[2 marks]

_____

_____

_____

> **! Exam Tip**
>
> For this question, you need to give your opinion and the reason you have that opinion.

01.2 The results of the monitoring are shown in **Figure 1**.

Figure 1

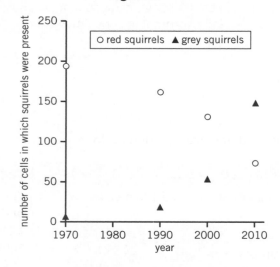

> **! Exam Tip**
>
> Draw a line for the red squirrels and another for the grey squirrels to make the pattern more obvious.

Describe the trends shown in the graph. **[3 marks]**

_____

_____

_____

_____

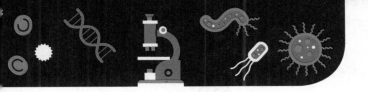
01.3 **Table 1** provides information on the two squirrel species.

Table 1

	Red squirrel	Grey squirrel
life expectancy	7 years	2–4 years
reproduction	up to six young, twice a year	up to nine young, twice a year
age of sexual maturity	12 months	12 months
survival rate of offspring	15 %	40 %
health	*Parapox* virus results in high levels of fatalities	carriers of *Parapox* virus

Using **Table 1**, suggest reasons for the changes seen in squirrel populations in Italy. **[4 marks]**

_____

_____

_____

_____

_____

01.4 Five breeding pairs of red squirrels are introduced into a new area, where no squirrel populations currently exist. Using data from **Table 1**, estimate the maximum number of offspring that survive from the breeding pairs after four years. **[4 marks]**

_____

_____

_____

_____

> **! Exam Tip**
>
> Take this year by year – how many offspring are there by the end of the first year, how many by the end of the second year.

01.5 Suggest reasons why the total population of red squirrels may be higher, or lower, than the value calculated in **01.4**. **[3 marks]**

_____

_____

_____

_____

02   Common seals are found in the seas surrounding the UK.

Figure 2

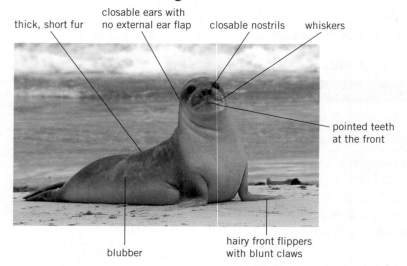

thick, short fur — closable ears with no external ear flap — closable nostrils — whiskers — pointed teeth at the front — blubber — hairy front flippers with blunt claws

02.1   Using **Figure 2** and your own knowledge, explain **two** ways seals are adapted for swimming.   **[4 marks]**

1 _____

_____

2 _____

_____

Exam Tip

Even if you haven't studied seals in class, you can apply your knowledge of adaptations in other animals to answer this question.

02.2   Suggest why seals have closable ears and nostrils.   **[1 mark]**

_____

_____

02.3   Arctic seals are another species of seal, which is adapted to live in cold regions of the world.

Suggest and explain **two** ways in which Arctic seals may differ from the common seal.   **[4 marks]**

1 _____

_____

2 _____

_____

03   An ecosystem is made up of a community interacting with the environment. The size and make-up of the community are affected by the abiotic and biotic factors present in the ecosystem. One example of an ecosystem is a woodland.

03.1   Describe what is meant by a community.   **[1 mark]**

03.2   Give **one** example of an abiotic factor in a woodland.   **[1 mark]**

03.3 Give **one** example of a biotic factor in a woodland. **[1 mark]**

03.4 Beech trees and ivy compete in a woodland ecosystem. Name **two** factors these species compete for. **[2 marks]**

03.5 Describe **one** way in which beech trees support the presence of animals in a woodland. **[1 mark]**

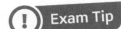

> **! Exam Tip**
> The biotic factors are the living ones.

04 Blackberry bushes are commonly found in UK hedgerows. They are often referred to as brambles as they have long, thorny, and arching stems, which can grow up to several metres tall.

04.1 Suggest **two** ways a blackberry bush is adapted to its environment. **[2 marks]**

04.2 Blackberry seeds are found within the blackberry fruit. Describe how blackberry seeds are dispersed. **[2 marks]**

04.3 Explain the advantages to the blackberry bush of dispersing its seeds. **[2 marks]**

> **! Exam Tip**
> Read the information in the question.

05 Within a community, each species depends on other species in order to survive. This is called interdependence.

05.1 Explain the interdependence that occurs between bees and cereal crops. **[4 marks]**

05.2 Fields of cereal are fairly unstable communities. Describe what is meant by a stable community. **[2 marks]**

05.3 Suggest and explain **two** ways a farmer could increase the stability of communities in their farmland, whilst still maximising crop yields. **[4 marks]**

> **! Exam Tip**
> For this question you have to say *why* you have made these suggestions.

06 Living organisms can be found almost everywhere on the planet.

06.1 Give the term that describes an organism that can live in an extreme environment. **[1 mark]**

06.2 Some species of fish are adapted to live at the bottom of the ocean. One example is the angler fish. Suggest **two** conditions that can be found at the bottom of the ocean. **[2 marks]**

06.3 The angler fish gets its name from an elongated spine that supports a light-producing organ known as a photophore. The photophore contains bacteria that can emit light through a chemical process known as bioluminescence. The photophore produces a blue-green light similar to that of a firefly on land. Give the type of relationship that exists between the bacteria and the angler fish. **[1 mark]**

06.4 Angler fish are blind. Suggest how their photophore helps them to survive. **[2 marks]**

> **! Exam Tip**
> Read the information in the question to help you work out this answer!

07 Marram grass (**Figure 3**) is adapted to live in very dry conditions, such as sand dune systems. The leaves of the marram grass are adapted to survive when water is limited. In very dry conditions, the leaves of the marram grass roll up to form long tubes. This helps drain any water down towards the roots of the plant.

Figure 3

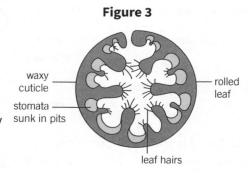

waxy cuticle

stomata sunk in pits

rolled leaf

leaf hairs

Using **Figure 3** and your own knowledge, explain how the leaves of marram grass are adapted to survive with very limited water availability. **[6 marks]**

08 Deer can be found living in the wild in large areas of woodland in the UK.

08.1 Which of the following describes a woodland in which deer live? Choose **one** answer. **[1 mark]**

a community   a population   an ecosystem   a habitat

08.2 Identify which of the following are examples of biotic factors that affect deer populations. Choose **two** answers. **[2 marks]**

hunting   light intensity   rainfall   mass of grass present

08.3 Intraspecific competition is competition within a species. Identify and explain **two** factors that deer compete with each other for. **[4 marks]**

08.4 Some factors can affect a population indirectly. For example, temperature can indirectly affect the population of starfish. This is because rising temperatures kill coral, which is the food source of the starfish. Explain how light intensity indirectly affects the size of a deer population. **[3 marks]**

09 The zebra mussel is a small mussel originally native to the Caspian Sea. A small population of these mussels was transferred to North America in the mid 1980s. One of the largest colonies exists in the Hudson River area. By the early 1990s, the biomass of the zebra mussel population exceeded the combined biomass of all other consumers in the Hudson River area. Mussels are filter feeders; this means that they filter small organisms and organic particles out of the water. This has significantly increased the river water clarity.

Suggest and explain the positive and negative effects of the invasive species on the native populations of organisms in the Hudson River area. **[6 marks]**

10 Desert foxes and Arctic foxes show both similarities and differences in their appearance.

These are summarised in **Table 2**.

Table 2

	Desert fox	Arctic fox
habitat	desert	ice sheets
fur colour	pale yellow	white
ears	large	small
feet	covered in hairs on both surfaces	covered in hairs on both surfaces
body features	specialised kidneys that reduce water loss from the body	have a thick layer of fat underneath the skin
average height of males	20 cm	30 cm
average mass of males	1.0 kg	6.0 kg

10.1 Explain why the two species of fox are different colours. **[2 marks]**

10.2 Suggest **two** reasons why the feet of the Arctic fox are covered in hairs. **[2 marks]**

10.3 The surface area-to-volume ratio of the foxes can be estimated by modelling their shape as a cube. Using this approach, the desert fox has a surface area-to-volume ratio of 3:10.

Estimate the surface area-to-volume ratio of an Arctic fox. **[3 marks]**

10.4 Explain why the two species of fox have different surface area-to-volume ratios. **[3 marks]**

11 A scientist carried out an investigation to study the effect of pH on enzyme action.

11.1 Give the type of enzyme that catalyses the breakdown of protein. **[1 mark]**

11.2 Give the product that is formed when protein is broken down. **[1 mark]**

> **! Exam Tip**
>
> Think about where each species lives.

> **! Exam Tips**
>
> For the desert fox we can assume each length is 20 cm.
>
> So the volume would be $20 \times 20 \times 20 = 8000 \text{ cm}^3$, with a surface area of $6 \times (20 \times 20) = 2400$.

> **! Exam Tip**
>
> Enzymes generally sound like the thing they are breaking down.

11.3 The scientist's results are shown in **Table 3**.

Table 3

pH	Mean rate of product formed in mmol/min
1	18
2	24
3	11
4	3
5	0

Plot a graph of the scientist's results on **Figure 4**. **[4 marks]**

Figure 4

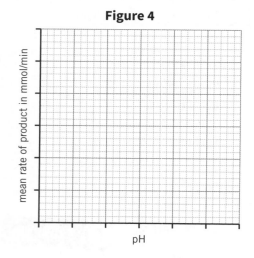

11.4 Use your graph to determine the optimum pH of the enzyme. **[1 mark]**

11.5 Explain what has happened to the enzyme at pH 5. **[3 marks]**

11.6 Suggest where in the body the enzyme is secreted. **[1 mark]**

12 Scientists classify organisms into groups.

12.1 Give **one** structure found inside a cell that scientists can use to classify organisms. **[1 mark]**

12.2 **Table 4** shows one way lions can be classified.

Table 4

Kingdom	Animalia
Phylum	Chordata
Class	Mammalia
	Carnivora
Family	Felidae
Genus	*Panthera*
Species	*leo*

Give the name of the scientist who devised this mechanism of classification. **[1 mark]**

12.3 Give the name of the missing category in **Table 4**. [1 mark]

12.4 Use **Table 4** to give the binomial name of lions. [1 mark]

12.5 Explain why classification systems are useful to scientists. [3 marks]

13 Scientists have used genetic engineering to create a new type of rice called golden rice. Scientists have added a gene to wild rice that makes it produce beta carotene. This changes the colour of the wild rice to a golden colour. Humans need beta carotene in order to make vitamin A, which is essential for good vision.

13.1 Explain an advantage of producing golden rice. [2 marks]

13.2 Describe how golden rice could be genetically engineered. [4 marks]

13.3 Suggest and explain **two** possible concerns people may have about eating golden rice. [4 marks]

> (!) **Exam Tip**
>
> Don't just think about the part of the world where you live, think about less economically-developed areas.

14 Yeast is a microorganism used in the brewing industry to make alcoholic drinks such as beer.

14.1 Write down the type of anaerobic respiration yeast carries out to produce ethanol. [1 mark]

14.2 Complete the word equation to describe this reaction. [2 marks]

_____ → ethanol + _____ (+ energy)

14.3 Describe how a scientist can test for the product you have named in **14.2**. [2 marks]

> (!) **Exam Tip**
>
> Knowing how to test for gases is important in all science subjects.

14.4 A group of students wanted to determine the optimum temperature for yeast to respire.

Suggest which piece of equipment they should use to control the temperature. [1 mark]

14.5 The students decided to measure the respiration rate at five different temperatures between 10 °C and 50 °C, at 10 °C intervals. Describe what you would expect the students to find as the temperature in the investigation was increased. [3 marks]

Knowledge

B19 Organising an ecosystem

Levels of organisation

Feeding relationships within a community can be represented by **food chains**.

Photosynthetic organisms that synthesise molecules are the producers of all **biomass** for life on Earth, and so are the first step in all food chains.

A range of experimental methods using transects and quadrats are used by ecologists to determine the distributions and abundances of different species in an ecosystem.

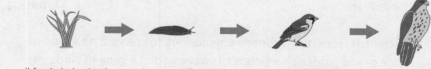

all food chains begin with a **producer**, for example, a green plant or alga producing glucose through photosynthesis

slugs are **primary consumers** – they are **herbivores** that eat producers

sparrows are **secondary consumers** – they are **carnivores** that eat primary consumers

hawks are **tertiary consumers** – they are carnivores that eat secondary consumers

Consumers that kill and eat other animals are predators, and those that are eaten are **prey**.

Apex **predators** are carnivores with no predators.

Organisms usually have more complex feeding relationships, with more than one predator or more than one food source. These can be shown in a **food web**.

Pyramids of biomass

The **trophic level** of an organism is the number of steps it is from the start of its food chain.

Pyramids of biomass represent the relative amount of biomass at each trophic level of a food chain.

Biomass is the amount of living or recently dead biological matter in an area. Biomass is transferred from each trophic level to the level above it in the food chain.

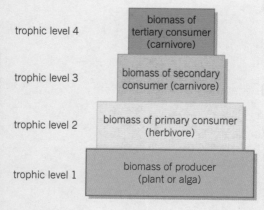

trophic level 4	biomass of tertiary consumer (carnivore)
trophic level 3	biomass of secondary consumer (carnivore)
trophic level 2	biomass of primary consumer (herbivore)
trophic level 1	biomass of producer (plant or alga)

Producers transfer about 1% of the incident light energy used for photosynthesis to produce biomass.

Approximately 10% of the biomass from each trophic level is transferred to the level above it.

How materials are cycled

All materials in the living world are recycled, which provides the building materials for future organisms.

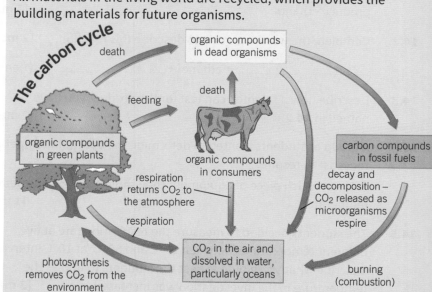

The carbon cycle

death

organic compounds in dead organisms

death

feeding

organic compounds in green plants

organic compounds in consumers

carbon compounds in fossil fuels

respiration returns CO₂ to the atmosphere

decay and decomposition – CO₂ released as microorganisms respire

respiration

CO₂ in the air and dissolved in water, particularly oceans

photosynthesis removes CO₂ from the environment

burning (combustion)

This loss of biomass moving up the food chain is due to several factors:

- use in life processes, such as respiration
- not all of the matter eaten is digested, some is egested as waste products
- some absorbed material is lost as waste
- energy is used in movement and to keep animals warm.

Key terms

Make sure you can write a definition for these key terms.

biomass   carbon cycle   carnivore
consumer   decomposer
evaporation   fertiliser   food chain
food web   herbivore   precipitation
predator   prey   producer
trophic level   water cycle

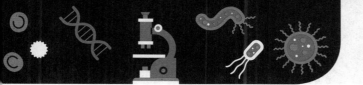

The water cycle

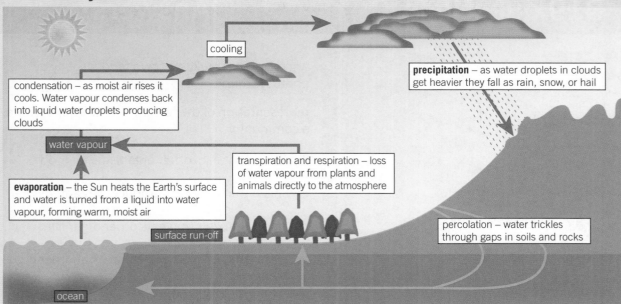

cooling

precipitation – as water droplets in clouds get heavier they fall as rain, snow, or hail

condensation – as moist air rises it cools. Water vapour condenses back into liquid water droplets producing clouds

water vapour

evaporation – the Sun heats the Earth's surface and water is turned from a liquid into water vapour, forming warm, moist air

transpiration and respiration – loss of water vapour from plants and animals directly to the atmosphere

surface run-off

percolation – water trickles through gaps in soils and rocks

ocean

Decomposition

Decomposers, such as bacteria and fungi, break down dead plant and animal matter by secreting enzymes into the environment. The small soluble food molecules produced then diffuse into the decomposer.

These materials are cycled through an ecosystem by decomposers returning carbon to the atmosphere as carbon dioxide and mineral ions to the soil.

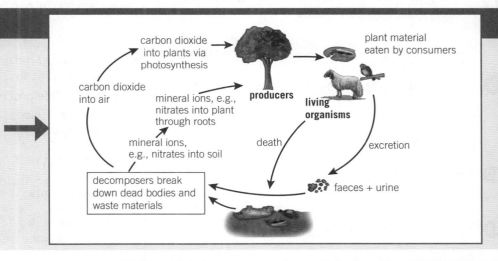

carbon dioxide into plants via photosynthesis

plant material eaten by consumers

carbon dioxide into air

mineral ions, e.g., nitrates into plant through roots

producers

living organisms

mineral ions, e.g., nitrates into soil

death

excretion

decomposers break down dead bodies and waste materials

faeces + urine

Gardeners and farmers try to provide optimum conditions for the rapid decay of waste material by decomposers.

Decomposition will occur faster in warm temperatures, when oxygen and moisture levels are high, and there is a neutral pH.

The compost produced from this decay is then added to soil as a natural **fertiliser** for growing garden plants and crops.

When there is a lack of oxygen, waste is decomposed anaerobically.

Anaerobic decay produces methane gas. Biogas generators use anaerobic decay to produce methane for use as a fuel.

Impacts of environmental change

Environmental changes affect the distribution of species in ecosystems.

These changes may be seasonal, geographic, or caused by humans, and include

- temperature – varies greatly between locations and seasons, and warming temperatures have contributed to species migrating away from the Equator
- availability of water – during droughts animals have to move away from their usual habitats to areas with more water, and cannot survive if this is not possible
- composition of atmospheric gases – human activities release greenhouse gases and pollutants, which cause harmful effects like climate change and acid rain.

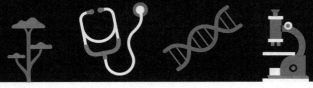

Learn the answers to the questions below, then cover the answers column with a piece of paper and write as many as you can. Check and repeat.

B19 questions	Answers
1 What is a producer?	organism that makes its own food, usually by photosynthesis
2 What is a food chain?	representation of the feeding relationships within a community
3 What is a consumer?	organism that eats other organisms for food
4 What is a herbivore?	organism that only eats producers (plants/algae)
5 What is a predator?	organism that kills and eats other organisms
6 What is a prey organism?	organism that is killed and eaten by another organism
7 What is an apex predator?	carnivore with no predators
8 What proportion of biomass is transferred from each trophic level to the one above?	approximately 10%
9 Why is biomass lost between trophic levels?	• some ingested material is egested • some material is lost as waste (carbon dioxide and water in respiration, water and urea in urine) • used in life processes, such as respiration • energy is used in movement and to keep animals warm
10 What is the carbon cycle?	process that returns carbon from organisms to the atmosphere as carbon dioxide, which can then be used by plants
11 What is the water cycle?	process that provides fresh water for plants and animals on land before draining into seas and rivers
12 What is a decomposer?	organism that breaks down dead plant and animal matter
13 What is the role of decomposition?	returns carbon to the atmosphere and mineral ions to the soil from dead matter
14 What factors affect the rate of decay by decomposers?	oxygen levels, moisture levels, temperature, and pH
15 What gas does anaerobic decay produce?	methane gas
16 How can this gas be used?	as a fuel
17 Give the environmental changes that can affect the distribution of organisms.	temperature, availability of water, and composition of atmospheric gases

The dotted divider column is labelled: Put paper here

Now go back and use the questions below to check your knowledge from previous chapters.

Previous questions — Answers

Previous questions	Answers
Name two culture media that microorganisms can be grown in.	nutrient broth solution, agar gel plates
What is the function of adrenaline in the body?	increases heart rate and boosts delivery of oxygen and glucose to brain and muscles to prepare the body for 'fight or flight'
List the abiotic factors that can affect a community.	• carbon dioxide levels for plants • light intensity • moisture levels • oxygen levels for animals that live in water • soil pH and mineral content • temperature • wind intensity and direction
Define the term heterozygous.	two different alleles present in an organism
What is evolution?	change in the inherited characteristics of a population over time through natural selection, which may result in a new species

Put paper here

Required Practical Skills

Practise answering questions on the required practicals using the example below. You need to be able to apply your skills and knowledge to other practicals too.

Rate of decay

For this practical, you need to be able to plan and carry out an experiment to explore how changing the temperature changes the rate of decay of milk.

To carry out this investigation, you must also be able to make accurate measurements of time and temperature, and use this to interpret a biological change. You should also be able to describe how an indicator solution can be used to measure the rate of a reaction.

Worked example

A student collected data on the time taken for milk mixed with Cresol red solution to turn from purple to yellow after adding lipase, at different temperatures.

Temperature in °C	Time taken in s
45	16
46	16
47	17
48	17
49	16

Evaluate the design of the experiment.

Answer: The range in the independent variable is 45–49 °C, and the interval is 1 °C. This is not a wide enough range or interval to give a change in the dependent variable. This is supported by the data – the time taken is the same at 45 °C as at 49 °C, and the results only change by a maximum of 1 second. A bigger range in the independent variable would give more reliable results.

Practice

1 Suggest a more accurate way to measure the colour change of the milk in the investigation.

2 The decay of milk is too slow to observe in a single lesson, so lipase is added to the milk to mimic natural decay. Identify the substance that the lipase breaks down, and give the name of the product that causes the milk pH to change.

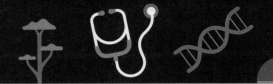
Exam-style questions

01 Isle Royale is a large isolated island on Lake Superior, in the USA. It is home to populations of wolves and moose. The wolves are the only natural predators of the moose.

The populations of wolves and moose on the island have been monitored over a 60-year period (**Figure 1**).

Figure 1

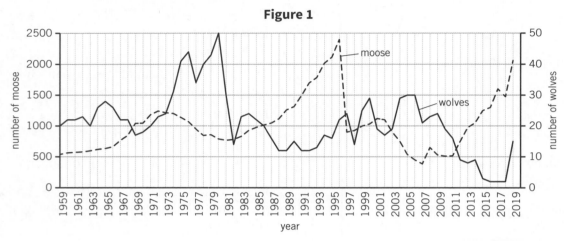

01.1 Identify the largest population of wolves on the island between 1959 and 2019. **[1 mark]**

_____

01.2 The population of moose rose significantly between 1982 and 1996. Calculate the percentage increase in the moose population between 1982 and 1996. **[3 marks]**

_____%

01.3 Explain the general population size trends in **Figure 1**. **[4 marks]**

_____

_____

_____

_____

01.4 Other than availability of moose, suggest and explain **two** factors that may affect the wolf population. **[2 marks]**

1 _____

_____

2 _____

_____

02 A group of students were asked to survey the number of daisies growing on their school field. They decided to take a series of samples using a 1 m × 1 m quadrat. Their results are in **Table 1**.

Table 1

Sample number	1	2	3	4	5	6	7	8
Number of daisies	11	14	14	2	14	15	11	15

02.1 Write down the median number of daisies from the students' samples. **[1 mark]**

_____

! Exam Tip

Rewrite the data to help you find the median.

02.2 Calculate the mean number of daisies from the students' samples. **[1 mark]**

! Exam Tip

Whenever you asked to find the mean, start by looking for anomalous results.

Mean = _____

02.3 Justify which value, from the median and the mean, gives the best estimate of the average number of daisies in the school field. **[2 marks]**

_____

_____

_____

02.4 The school field measures 350 m × 200 m.

Estimate the number of daisies in the school field. **[3 marks]**

Number = _____

02.5 Suggest and explain **two** reasons for the differences in number of daisies measured in Sample **4**. **[4 marks]**

1 _____

_____

2 _____

_____

! Exam Tip

There is lots of information that is not specified in the main body of the question. This leaves you lots of space to suggest reason. As long as you've got a good explanation you can easily pick up marks in **02.5**.

03   A group of students wanted to estimate the number of thistle plants growing on a large lawn.

03.1   Which biotic factor could affect the population of thistles?
Choose **one** answer.   **[1 mark]**

light intensity   presence of daisies

soil pH   water availability

03.2   Identify the piece of equipment they should use in their investigation.   **[1 mark]**

pooter   quadrat   stopwatch   tape measure

03.3   Describe the procedure the students should follow to estimate the number of thistle plants present in the lawn.   **[4 marks]**

03.4   Explain **one** step the students should take to ensure their estimated value is as accurate as possible.   **[2 marks]**

! Exam Tip

This is required practical, but in a slightly different context.

04   **Figure 2** represents the main steps in the carbon cycle.

Figure 2

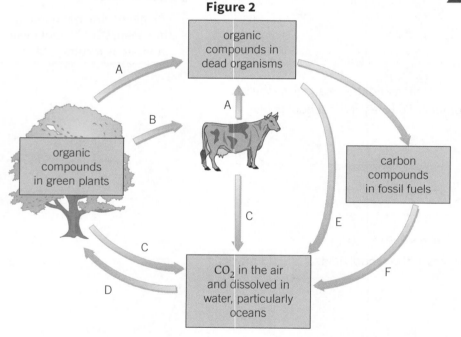

04.1   Write down the letter of the arrow that represents photosynthesis.
[1 mark]

04.2   Name the process that arrow **F** represents.   **[1 mark]**

04.3   Carbon is trapped inside the bodies of organisms. Name **one** other store of carbon.   **[1 mark]**

04.4   Describe the role that microorganisms play in the carbon cycle.
[3 marks]

! Exam Tip

Questions **04.1**, **04.2**, and **04.3** are only worth 1 mark. You shouldn't write more than one letter or two words for each answer. Rewriting the question will only waste time and not get you any more marks.

05 **Figure 3** is a pyramid of biomass that represents a food chain in a woodland.

Figure 3

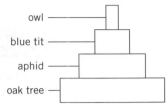

owl
blue tit
aphid
oak tree

05.1 Name the organisms that are found in trophic level 3. **[1 mark]**

05.2 Describe the group of organisms that are found in trophic level 1 of any food chain. **[1 mark]**

05.3 Identify the apex predator. **[1 mark]**

05.4 Decomposers are missing from **Figure 3**. Explain the role that decomposers play in food chains. **[3 marks]**

06 When milk decays, bacteria convert lactose into lactic acid. A group of students investigated how temperature affects the rate of decay of milk. Milk may take several days to decay, so the students used the following method to model the process:

1 Prepare five water baths: 20 °C, 30 °C, 40 °C, 50 °C, and 60 °C.

2 Place 20 cm³ alkaline solutions of milk into five separate conical flasks.

3 Place one conical flask in each water bath and leave to acclimatise.

4 Add a few drops of cresol red indicator to each beaker. Cresol red turns yellow when the pH drops below 7.2.

5 Add 2 cm³ lipase to each beaker.

6 Time how long it takes for a colour change to take place in each beaker.

7 Repeat the entire experiment.

The students' results are shown in **Table 2**.

Table 2

Temperature in °C	Time until colour change in s			Rate of reaction in _____
	Test 1	Test 2	Mean	
20	240	280	260	0.004
30	64	58	61	0.016
40	38	44	41	
50	120	100	110	0.009
60	did not change	did not change	–	–

06.1 Complete the table heading by adding the correct unit to the rate of reaction column. **[1 mark]**

06.2 Calculate the rate of reaction at 40 °C. **[2 marks]**

06.3 Plot a graph of temperature versus rate of reaction. Draw a line of best fit. **[3 marks]**

Figure 4

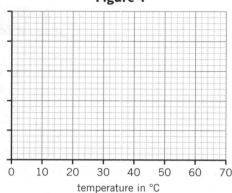

temperature in °C

06.4 Determine the optimum temperature for decay. **[2 marks]**

06.5 Explain why the students were unable to gain a result at a temperature of 60 °C. **[2 marks]**

06.6 Explain why the apparatus provided to the students provides a model for the decay of milk. **[6 marks]**

07 **Figure 5** shows a food chain of some organisms found in a park.

Figure 5

oak tree → caterpillar → sparrow → hawk

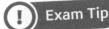

07.1 Identify the producer in this food chain. **[1 mark]**

07.2 Identify the tertiary consumer in this food chain. **[1 mark]**

07.3 Give the source of energy for this food chain. **[1 mark]**

07.4 Identify **one** prey organism from this food chain. **[1 mark]**

07.5 Suggest and explain **one** adaptation of the prey organism named in **07.4**. **[2 marks]**

07.6 Suggest and explain what would happen to the number of caterpillars if the hawk population was infected with a fatal virus. **[3 marks]**

08 All materials in the living world are recycled to provide the building blocks for future organisms. Water is one example of a material that is recycled.

08.1 Describe the main steps in the water cycle. **[6 marks]**

08.2 Identify and explain **two** ways in which animals return water to the environment. **[4 marks]**

08.3 Explain the importance of the water cycle to living organisms. **[4 marks]**

08.4 Other than water, name **one** material that is recycled. **[1 mark]**

> **Exam Tip**
>
> Always draw your graph using crosses to plot the points and add on a line of best fit.

> **Exam Tip**
>
> There is only one source of energy for all food chains.

> **Exam Tip**
>
> You can gain lots of marks from carefully annotated diagrams. Make sure you label each arrow in full sentences and don't spend ages on drawings – boxes with words in work just as well.

09   A group of students investigated the plants and animals present in a city park. They completed a line transect over 10 m to one side of a footpath. The number of organisms at each point was noted. They presented their data on a kite diagram (**Figure 6**).

Figure 6

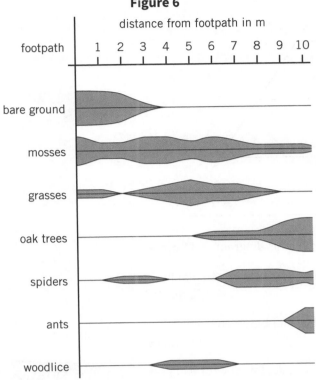

09.1   Describe a procedure the students could have followed to collect these data.   **[6 marks]**

09.2   Identify which animal species had the highest population. **[1 mark]**

09.3   Write down **two** conclusions the students could draw from their data.   **[2 marks]**

09.4   Suggest and explain **one** way the students could gather further evidence to support their conclusions.   **[2 marks]**

10   The carbon cycle is essential for life on Earth.

10.1   Describe the main processes in the carbon cycle.   **[6 marks]**

10.2   Explain how **two** human activities are causing changes to the natural balance of the carbon cycle.   **[4 marks]**

11   **Figure 7** shows the biomass of organisms required at each level of a food chain to support the next trophic level.

Figure 7

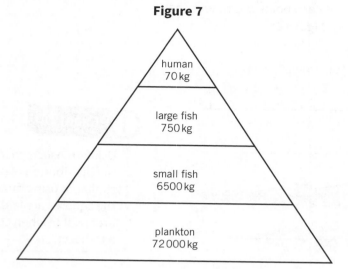

11.1   Write down the food chain represented by this diagram.   **[1 mark]**

11.2   Calculate the percentage of biomass transferred between plankton and humans.   **[2 marks]**

11.3   Explain why the efficiency of biomass transfer between trophic levels is much less than 100 %.   **[2 marks]**

11.4   The energy content of small fish is 6 kJ/g. Calculate the total energy transferred to the large fish in the food chain.   **[5 marks]**

11.5   Lead is a metal linked to liver and brain damage in humans. Some industrial processes produce lead as a by-product, which can contaminate water sources. Using **Figure 7**, suggest how eating a diet rich in fish could cause lead-related health disorders in humans.   **[4 marks]**

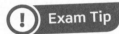

! **Exam Tip**

You will need to carefully show your working. This will help you pick up marks even if you don't get the final answer correct.

12   The populations of lynx and snowshoe hare in Canada were studied over a number of years. The data collected are summarised in **Table 3**.

Table 3

Year	Hare numbers	Lynx numbers
1895	85 000	48 000
1900	18 000	6000
1903	65 000	18 000
1905	40 000	61 000
1908	28 000	28 000
1909	25 000	4000
1910	51 000	10 000
1912	70 000	32 000
1915	30 000	42 000

12.1 Using **Table 3**, describe the relationship between the snowshoe hare and lynx populations. **[2 marks]**

Exam Tip

You will have to include data from **Table 3** in your answer.

12.2 Explain whether the data provide evidence for a stable community. **[4 marks]**

12.3 Calculate the percentage change in the lynx population between 1909 and 1912. **[3 marks]**

12.4 Suggest and explain **two** factors that may be responsible for the unusually large hare population in 1895. **[2 marks]**

12.5 Suggest and explain what would happen to the forest environment if lynx started being hunted for their fur. **[3 marks]**

13 If you touch a hot saucepan, your body responds rapidly using a reflex response.

13.1 Identify the correct order of neurones in a reflex arc.
Choose **one** answer. **[1 mark]**

stimulus → sensory neurone → relay neurone → motor neurone → effector

stimulus → motor neurone → relay neurone → sensory neurone → effector

stimulus → sensory neurone → motor neurone → relay neurone → effector

stimulus → relay neurone → sensory neurone → motor neurone → effector

13.2 The average speed at which an impulse travels through this pathway is 65.5 m/s. The length of the pathway from the receptor cell in a person's finger to the muscle in the arm is 1.4 m. Calculate the time it takes for the impulse to travel along this reflex arc. **[2 marks]**

Exam Tip

There are no equations that include this specifically for biology but you are expected to have a good general knowledge of the equations you need for maths.

13.3 Explain **two** ways a sensory neurone is adapted to its function. **[2 marks]**

13.4 Coordination in the body can be brought about by nervous or hormonal responses. Discuss the similarities and differences between nervous and hormonal coordination in the body. **[4 marks]**

14 Genetic information is passed on to an organism's offspring through its DNA.

14.1 Explain why children of the same parents look similar but not identical. **[4 marks]**

Exam Tip

This question is about siblings not only about twins.

14.2 Write down what is meant by a genetic mutation. **[1 mark]**

14.3 Give **one** possible advantage and **one** possible disadvantage of a genetic mutation. **[2 marks]**

Knowledge

B20 Humans and biodiversity

Biodiversity

Biodiversity is the variety of all the different species of organisms (plant, animal, and microorganism) on Earth, or within a specific ecosystem.

High biodiversity ensures the stability of an ecosystem, because it reduces the dependence of one species on another in the ecosystem for food or habitat maintenance.

The future of the human species depends on us maintaining a good level of biodiversity. Many human activities, such as **deforestation**, are reducing biodiversity, but only recently have measures been taken to try to prevent this.

Waste management

Rapid growth of the human population and increases in the standard of living mean humans are using more resources and producing more waste.

Waste and chemical materials need to be properly handled in order to reduce the amount of **pollution** they cause. Pollution kills plants and animals, and can accumulate in food chains, reducing biodiversity.

Pollution can occur

- in water, from sewage, fertiliser run-off, or toxic chemicals (e.g., from factories)
- in air, from smoke and acidic gases
- on land, from landfill and toxic chemicals.

Maintaining biodiversity

Many habitats are currently under threat due to human activities such as deforestation, climate change, and habitat destruction.

There are a number of ways in which scientists and concerned citizens are trying to maintain biodiversity and reduce the negative impact of humans on ecosystems, including

- breeding programmes in zoos for endangered species
- protection and regeneration of rare habitats (e.g., as national parks)
- reintroduction of hedgerows in agricultural areas where single crop species are grown, as hedges provide habitat for many organisms
- government policies to reduce deforestation and carbon dioxide emissions
- recycling resources rather than dumping waste in landfill.

Land use

Rapid population growth has led to humans using much more land for building, quarrying, farming, and dumping waste. This reduces the area in which animals can live and can further destroy habitats through pollution.

For example, the destruction of **peat bogs** (areas of partially decayed vegetation) to produce garden compost has decreased the amount of this important habitat, and the biodiversity it supports. The decay or burning of peat for energy also releases carbon dioxide into the atmosphere, contributing to **global warming**.

Deforestation

Large-scale deforestation in tropical areas has been carried out to provide land for cattle and rice fields, and to grow crops for **biofuels**.

This has resulted in

- large amounts of carbon dioxide being released into the atmosphere due to burning of trees
- extinctions and reductions in biodiversity as habitats are destroyed
- climate changes, as trees absorb carbon dioxide and release water vapour.

 **Key terms**

Make sure you can write a definition for these key terms.

| biodiversity | biofuel | deforestation | food security |
| global warming | intensive farming | peat bog | pollution |

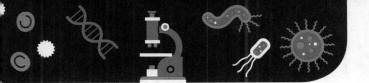

Food security

Food security is having enough food to feed a population.

Biological factors threatening human food security include

- rapid population growth and increasing birth rate in some countries
- changing diets in developed countries, requiring scarce food resources to be transported globally
- new pests and pathogens impacting farming of vast amounts of crops
- environmental changes, such as drought, affecting food production
- increasing cost of agricultural inputs, like fertilisers
- conflicts in some parts of the world, which affect the availability of water or food.

Farming techniques

Sustainable methods of food production need to be developed if we are going to feed the Earth's human population.

Intensive farming techniques make food production more efficient by restricting energy transfer from food animals to their environment.

This can be done by

- limiting the movement of the animals
- controlling the temperature of their surroundings.

In order to also maximise yield from animals and crops, farmers also

- feed animals high-protein foods to increase growth
- give animals antibiotics to prevent or treat disease
- regularly use fertilisers, herbicides, and pesticides on crops.

Sustainable fisheries

Fish stocks in the oceans are declining. It is important to maintain fish stocks to ensure breeding continues, or certain species may disappear altogether in some areas.

To avoid this happening, net sizes (bigger holes to stop young fish being caught) and fishing quotas (how many fish can be caught) are controlled in many places.

Role of biotechnology

Modern biotechnology techniques enable increased food production to feed and maintain the rapidly increasing human population:

- Large quantities of microorganisms can be cultured for food – for example, mycoprotein, a protein-rich vegetarian food harvested and purified after growing the fungus *Fusarium* on glucose syrup in aerobic conditions.
- Genetically modified (GM) crops can have increased yields, increased resistance to changes in their environments, or improved nutritional values (e.g., golden rice).
- Bacteria can be genetically modified to produce human insulin that can be harvested, purified, and used to treat diabetes.

Global warming

Levels of carbon dioxide and methane in the atmosphere are increasing due to human activity, contributing to global warming and climate change. Global warming is the gradual increase in the average temperature of the Earth.

This scientific consensus is based on systematic reviews of thousands of peer-reviewed publications.

Global warming has resulted in

- large-scale habitat change and reduction, causing decreases in biodiversity
- extreme weather and sea level changes
- migration of species to different parts of the world, affecting ecosystems
- threats to the security and availability of food.

Advantages of intensive farming	Disadvantages of intensive farming
• high yield and quicker growth of crops and animals • efficient use of food, with less waste produced • can meet demand for food from a rapidly increasing population	• increased risk of antibiotic-resistant bacteria strains • pesticides and herbicides may kill beneficial organisms and reduce biodiversity • ethical issues about animal welfare and quality of life • large carbon dioxide and methane emissions

Retrieval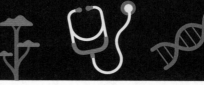

Learn the answers to the questions below, then cover the answers column with
a piece of paper and write as many as you can. Check and repeat.

B20 questions | Answers

	B20 questions	Answers
1	What is biodiversity?	the variety of all the different species of organisms on Earth, or within an ecosystem
2	What is the advantage of high biodiversity?	ensures stability of ecosystems by reducing the dependence of one species on another
3	How are humans trying to maintain biodiversity?	• breeding programmes • protection of rare habitats • reintroduction of hedgerows • reduction of deforestation and carbon dioxide emissions • recycling resources
4	Why are more resources being used, and more waste produced, by humans?	rapid growth in human population, and increase in the standard of living
5	Where does pollution occur?	water, air, and land
6	How are humans reducing the land available for other organisms?	building, quarrying, farming, and dumping waste
7	What are the negative impacts of the destruction of peat bogs?	• reduces amount of available habitat, causing decreases in biodiversity • burning or decay of peat releases carbon dioxide into the atmosphere
8	Why have humans carried out large-scale deforestation in tropical areas?	• provide land for cattle and rice fields • grow crops for biofuels
9	What is food security?	having enough food to feed a population
10	What are the biological factors threatening human food security?	• rapid population growth and increasing birth rate • new pests and pathogens • changing diets in developed countries • environmental changes • conflicts • costs of agricultural inputs
11	How can the efficiency of food production in farming be increased?	• limit movement of animals • control temperature of surroundings • feed animals high-protein foods • give animals antibiotics • regularly use fertilisers, herbicides, and pesticides
12	What gases are increasing in atmospheric levels and contributing to global warming?	carbon dioxide and methane
13	How can fish stocks be maintained at a sustainable level?	controlling net sizes and introducing fishing quotas
14	How is biotechnology used to maintain the growing human population?	• large quantities of microorganisms cultured for food, such as mycoprotein from *Fusarium* • GM bacteria producing treatments like human insulin • GM crops providing higher yields or improved nutritional values

Put paper here

Now go back and use the questions below to check your knowledge from previous chapters.

Previous questions — Answers

Previous questions		Answers
What type of cell division is involved in asexual reproduction?	Put paper here	mitosis
What is the function of thyroxine in the body?		stimulates basal metabolic rate, so is important for growth and development
What is a food chain?	Put paper here	representation of the feeding relationships within a community
What is sexual reproduction?		joining (fusion) of male and female gametes
Give the environmental changes that can affect the distribution of organisms.		temperature, availability of water, and composition of atmospheric gases
What evidence supports the theory of evolution?	Put paper here	• parents pass on their characteristics to offspring in genes • fossil record evidence • evolution of antibiotic-resistant bacteria
What type of cell division is involved in sexual reproduction?		meiosis
How do antibiotic-resistant strains of bacteria develop?		mutations that allow the strain to survive and reproduce

Required Practical Skills

Practise answering questions on the required practicals using the example below.
You need to be able to apply your skills and knowledge to other practicals too.

Field investigations

For this investigation, you will practise applying appropriate sampling techniques in the field to look at plant population sizes. Two methods of sampling with quadrats are covered:

1 Transect lines – stretching a tape measure along the ground, placing a quadrat at even points along the measure, and recording the number of plants within each quadrat.

2 Random sampling – using tape measures to form a square area, generating random numbers corresponding to where in that area you should place the quadrat, and recording the number of plants within each quadrat.

You should be able to describe and explain the purpose of each sampling method.

Worked example

A student used a quadrat measuring 25 cm by 25 cm to sample the number of daisies in a field. The average number of daisies within a quadrat was found to be 17. The total area of the field was 320 m².

1 Estimate the number of daisies in the field.

$$25\,cm = 0.25\,m$$
$$\text{area of quadrat} = 0.25 \times 0.25$$
$$= 0.0625\,m^2$$
$$\frac{320}{0.0625} = 5120$$
$$\text{population estimate} = 5120 \times 17$$
$$= 87\,040\,\text{daisies}$$

2 Give a reason why the student might use random sampling in this investigation.

Random sampling reduces any bias to the results, meaning they are more reliable.

Practice

1 A student wanted to measure how distance from a water source affected the size of a plant. Write a method to carry out this investigation.

2 A quadrat measures 15 cm by 15 cm. Give the area of the quadrat in m².

3 An ecologist wanted to estimate the quantity of plastic floating in the ocean. Write a method for this investigation.

Exam-style questions

01 Large areas of the Amazon rainforest have been cleared for agriculture.

01.1 Choose the word that best describes this change in land use.

[1 mark]

afforestation ☐

deforestation ☐

eutrophication ☐

leaching ☐

01.2 Give **three** ways in which the removal of trees can lead to increasing carbon dioxide levels in the Earth's atmosphere. **[3 marks]**

1 _____

2 _____

3 _____

> **! Exam Tip**
>
> In **01.2** you will be awarded 1 mark per way, so don't write long sentences explaining why.

01.3 Explain why there is a difference in the level of biodiversity between rainforest regions and agricultural land. **[3 marks]**

_____

_____

_____

_____

01.4 Explain the long-term effects on the Earth's atmosphere of this change in land use. **[4 marks]**

_____

_____

_____

_____

_____

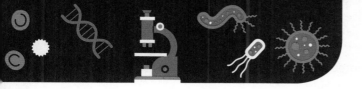

02 **Figure 1** shows how atmospheric carbon dioxide (CO_2) concentration and global mean temperature have changed since 1990.

Figure 1

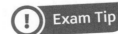

02.1 Describe the changes since 1900. **[4 marks]**

_____

_____

_____

_____

_____

! Exam Tip

You'll need to talk about both lines, give the general trend, and use data from the graph, for example, in 1900 the carbon dioxide concentration was … since then it has … and the level is now … .

02.2 Name **two** activities that have led to the changes described in **02.1**. **[2 marks]**

1 _____

2 _____

02.3 The global mean temperature in the twentieth century was 13.9 °C. Calculate the global mean temperature in the year 2000. **[1 mark]**

_____

Mean = _____

02.4 Calculate the mean rate of change of CO_2 concentration between the years 1910 and 2000. Include an appropriate unit with your answer. **[3 marks]**

Mean = _____ Unit = _____

! Exam Tip

To determine the units for rate of change, look at the units for the two values you are using and the calculation you did.

02.5 A student states that 'recent changes in global temperatures are caused by the changes in atmospheric carbon dioxide concentration'.

Discuss the extent to which **Figure 1** supports the student's statement. **[5 marks]**

_____

_____

_____

_____

_____

_____

_____

_____

_____

> **! Exam Tip**
>
> **02.5** refers to changes. You will need to be specific in your answer and give data from **Figure 1** to support what you say.

03 Air pollution is a significant problem in some UK cities.

03.1 Give **one** reason why air pollution is a concern for humans. **[1 mark]**

03.2 Describe what is meant by smog. **[1 mark]**

03.3 Nitrogen oxides (NO_x) are a type of air pollutant present in cities. Identify the main source of emission of NO_x. Choose **one** answer. **[1 mark]**

car exhausts   burning coal

burning wood   factories

> **! Exam Tip**
>
> NO_x is just referring to the different nitrous oxides that can have different numbers of oxygen in them.

03.4 **Table 1** shows how the concentration of NO_x varies over a one-week period in a UK city.

Table 1

NO_x concentration in µg/m³	125	110	120	123	130	85	70
Day	Monday	Tuesday	Wednesday	Thursday	Friday	Saturday	Sunday

Suggest **one** conclusion from these data. Give a reason for your answer. **[2 marks]**

03.5 Calculate the mean atmospheric concentration of NO_x. **[1 mark]**

03.6 Suggest and explain **one** way this figure could be reduced. **[2 marks]**

> **! Exam Tip**
>
> This data is from a city. To see any patterns in context think about the different activities that go on over a week and over a weekend.

04 Biodiversity is important in maintaining a stable ecosystem.

04.1 Define what is meant by the term biodiversity. **[1 mark]**

04.2 Identify the **two** factors that can increase biodiversity. **[2 marks]**

breeding programmes   deforestation

monoculture farming   reintroduction of hedgerows

04.3 Describe why an area of woodland with multiple species of tree is more stable than an area containing only one tree species.

[3 marks]

05 **Table 2** shows the percentage recycling rate in the UK over five years.

Table 2

Year	Recycling rate in %
2010	40.4
2011	42.9
2012	43.9
2013	44.1
2014	44.9

05.1 Name **two** materials that can be recycled. **[2 marks]**

05.2 Explain **two** environmental benefits of recycling. **[4 marks]**

05.3 The UK was set a target to recycle at least 50 % of household waste by 2020. Based on the data in **Table 2**, evaluate the likelihood that this target will be met. **[4 marks]**

(!) Exam Tip

You'll have to back up what you say with a calculation based on data from **Table 2**.

06 The total area of peat bogs and peatlands in the UK is decreasing.

06.1 Describe how peat is formed. **[2 marks]**

06.2 Describe **two** uses of peat. **[2 marks]**

06.3 Explain **two** reasons why the human use of peatlands causes negative effects on the environment. **[4 marks]**

06.4 To help reduce the human impact on the environment, the UK government is encouraging gardeners to use peat-free compost. Suggest an alternative source of compost. **[1 mark]**

(!) Exam Tip

Peat is a bit like mud and can be found in parts of Scotland. It can be dug out of the ground and left to dry before being used.

07 **Figure 2** shows the change in the human population since the year 1800.

Figure 2

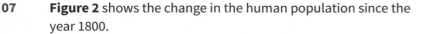
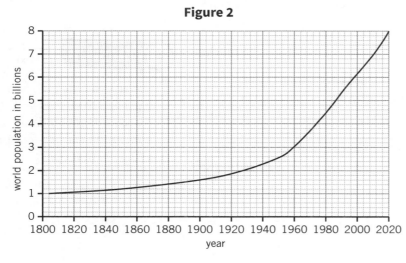

(!) Exam Tip

The y-axis has the population in billions. Be careful that you use the correct unit.

07.1 Describe the change in population in **Figure 2**. **[3 marks]**

07.2 Describe **three** ways the change in population has led to an increase in pollution. **[3 marks]**

07.3 Other than increases in pollution, suggest and explain **two** environmental issues that this change in human population could cause. **[4 marks]**

08 The article in **Figure 3** was published in an online magazine.

Figure 3

> **Stop biodiversity loss – or face human extinction**
>
> The United Nations' biodiversity chief warned today that the human race faces a biodiversity crisis. At a recent conference, it was reported that 'Ambitious plans are required for all countries to draw up plans to protect the insects, birds, and animals upon which the human race relies as raw materials for its survival. Without these materials the human race faces a potential mass extinction event.' Many prominent environmental groups were quick to support this position, one adding that 'Around 1000 species per year are becoming extinct. These species could be hugely beneficial to the human race – but now we will never know.'

Exam Tip

This is a large block of text. Read the questions first, then read the block of text. This way you can pick out key bits of information as you read the text. For example:

- **08.2** wants examples of human activity that decrease biodiversity, so highlight anything that decrease biodiversity in one colour
- **08.3** wants projects carried out by scientists, so highlight any project in another colour.

08.1 Define the term biodiversity. **[1 mark]**

08.2 Describe **two** examples of human activity that cause a decrease in biodiversity. **[2 marks]**

08.3 Give **two** projects carried out by scientists that aim to maintain or increase biodiversity. **[2 marks]**

08.4 Suggest and explain **two** ways in which a loss of biodiversity could lead to a mass extinction event for the human race. **[6 marks]**

09 One of the biggest challenges facing many less-developed countries is providing food security.

09.1 Explain what is meant by the term food security. **[1 mark]**

09.2 Explain **two** ways that food security may be reduced in a less-developed country. **[4 marks]**

Exam Tip

For an evaluate question you will need to give advantages, disadvantages, and your own conclusion with the reason you have that opinion.

09.3 Food security can be increased by the use of efficient farming techniques. Explain why the intensive farming of pigs makes the production of pork products more efficient. **[3 marks]**

09.4 Evaluate the advantages and disadvantages of intensive farming. **[4 marks]**

10 For many years fish stocks in the oceans have been declining as a result of overfishing. Measures have been introduced to make fishing more sustainable.

10.1 Describe the main aims of sustainable fishing. **[2 marks]**

10.2 Give **one** measure that can be used to make commercial fishing more sustainable. **[1 mark]**

10.3 Explain how controlling the size of holes in nets supports sustainable fishing. **[3 marks]**

11 To feed the world's growing population, scientists are developing new methods of food production. One example is the development of mycoprotein from fungus.

11.1 Describe **two** benefits of using mycoprotein as a food source. **[2 marks]**

11.2 Explain how mycoprotein is produced. **[4 marks]**

11.3 Under optimum conditions the fungus doubles its mass every five hours.

Describe **three** ways scientists provide optimum conditions for the fungus's growth. **[3 marks]**

11.4 Producing protein from fungus is much more efficient than producing it from meat: 1000 g of plant carbohydrate can produce up to 14 g of beef or 136 g of mycoprotein.

Calculate the percentage difference in efficiency. **[2 marks]**

12 A person's characteristics can be the result of environmental causes, genetic causes, or both.

12.1 For each characteristic, draw **one** line to the most appropriate cause for the characteristic. **[2 marks]**

Characteristic		**Cause**
body mass		genetic only
blood group		environmental only
presence of tattoos		both

> ! **Exam Tip**
>
> You don't have to start with body mass if you're not sure of the answer.

For answers and more practice questions visit www.oxfordrevise.com/scienceanswers
Even more practice and interactive revision quizzes are available on **kerboodle**
B20 Practice 241

12.2 When a characteristic is inherited, two versions of a gene are passed on from parents to their offspring.

Write down the name given to the two versions of an inherited gene. **[1 mark]**

12.3 The two forms of an inherited gene are called dominant and recessive.

Explain why characteristics linked to recessive gene variants are less likely to be displayed. **[2 marks]**

12.4 The gene variant for right-handedness (R) is dominant, over left-handedness (r).

Complete the Punnett square to show that there is a 25 % chance of a child inheriting left-handedness from their parents. **[1 mark]**

! Exam Tip

When completing a Punnet square, full one coloumn or one row, rather than box by box. You're less likely to make mistakes that way.

		Father's gametes	
		R	r
Mother's gametes	R		
	r		

12.5 Explain why the child of a mother and father who are both left-handed will always be left-handed. **[3 marks]**

13 Controlling blood glucose levels in the human body is an example of homeostasis.

13.1 Define the term homeostasis. **[1 mark]**

13.2 The flow chart in **Figure 4** shows how the body responds when the blood glucose level becomes too high or too low.

! Exam Tip

These words are all very similar so take extra care. Spelling could be the key to showing the examiner what you know and them not being sure you can show the difference between two substances.

Figure 4

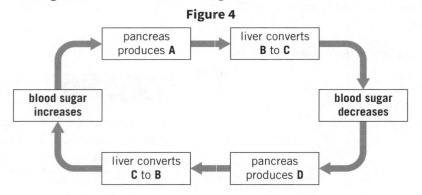

Name substances **A**, **B**, **C**, and **D**. **[4 marks]**

13.3 A diabetic and a non-diabetic volunteer took part in a study of blood glucose levels. **Table 3** shows how their blood glucose levels varied over time. They both ate breakfast at 08:00.

Table 3

| Time | Blood sugar levels in mg per litre of blood | |
	Person A	Person B
07:00	80	75
08:00	75	70
09:00	135	340
10:00	100	280
11:00	85	235
12:00	75	200

Identify and explain which volunteer was diabetic and which volunteer was non-diabetic. **[3 marks]**

13.4 Calculate the maximum rate of change of blood glucose concentration for Person **B** in mg per litre per minute. **[3 marks]**

13.5 Estimate the time at which Person **B**'s blood glucose concentration will return to the pre-meal level. **[2 marks]**

14 Gorse plants have a number of adaptations to help them survive.

14.1 Draw **one** line between each adaptation and its function. **[3 marks]**

Adaptation	Function
flowers	reduce water loss
sharp spines	so herbivores do not eat the plant
small leaves	to attract insects to pollinate them

14.2 Gorse plants grow in sunny regions, usually in dry sandy soils. Suggest and explain how the roots of gorse plants may appear. **[2 marks]**

14.3 Other than water, give **two** factors that plants compete for. **[2 marks]**

14.4 Apart from competition with other gorse plants, a number of biotic factors can affect the population of gorse plants present in an area. Describe how **one** other biotic factor could reduce the population of gorse bushes in an area. **[2 marks]**

(!) Exam Tip

The question has already given water, so you won't get any marks if you say that again.

OXFORD
UNIVERSITY PRESS

Great Clarendon Street, Oxford, OX2 6DP, United Kingdom

Oxford University Press is a department of the University of Oxford.

It furthers the University's objective of excellence in research, scholarship, and education by publishing worldwide. Oxford is a registered trade mark of Oxford University Press in the UK and in

certain other countries

British Library Cataloguing in Publication Data

Data available

978-1-38-200484-8

10 9 8 7 6 5

Paper used in the production of this book is a natural, recyclable product made from wood grown in sustainable forests.

The manufacturing process conforms to the environmental regulations of the country of origin.

Printed in Great Britain by Bell and Bain Ltd, Glasgow

Acknowledgements

The publisher would like to thank the following for permissions to use copyright material:

Jo Locke would like to thank Dave, Emily and Hermione for their support, as well as providing plenty of tea and cake.

Jessica Walmsley would like to give special thanks to her husband Joe, mum Barbara and Dad Dean for their continued support in everything she does.

Cover illustration: Andrew Groves

p7: JACK BOSTRACK, VISUALS UNLIMITED/SCIENCE PHOTO LIBRARY;
p8, **28**, **94**, **33**, **58**, **129**, **177**, **209**, **214**, **217**, **217**: Shutterstock;
p69: POWER AND SYRED/SCIENCE PHOTO LIBRARY.

Other artwork by Q2A Media Services Inc. and OUP.

Although we have made every effort to trace and contact all copyright holders before publication this has not been possible in all cases. If notified, the publisher will rectify any errors or omissions at the earliest opportunity.